HYDROGELS

Edited by **Sajjad Haider** and **Adnan Haider**

Hydrogels
http://dx.doi.org/10.5772/intechopen.68817
Edited by Sajjad Haider and Adnan Haider

Contributors

Ángel Serrano-Aroca, Leila Figueiredo Miranda, Kátia Lúcia Cunha, Antonio Munhoz Júnior, Isabella Tereza Barbosa, Terezinha Masson, Mohamed Mohamady Ghobashy, D Darminto, Malik Anjelh Baqiya, S Sunaryono, M Munaji, Dita Puspita Sari, Yanurita Dwi Hapsari, Oleh Suberlyak, Volodymyr Yosyp Skorokhoda, Carlo Luca Romano', Gaetano Giammona, Giovanna Pitarresi, Fabio Salvatore Palumbo, Susanna Maraldi, Sara Scarponi, Ailton De Souza Gomes, Fernanda Garcia Cordeiro Tessarolli, Claudia Regina Elias Mansur, Javad Foroughi, Azadeh Mirabedini, Holly Warren, Birzabith Mendoza-Novelo, Jesús Alejandro Claudio-Rizo, Jorge Delgado, Iraís Quintero-Ortega, José Mata-Mata

First published in London, United Kingdom, 2018 by IntechOpen
IntechOpen is the global imprint of INTECHOPEN LIMITED, registered in England and Wales, registration number: 11086078, The Shard, 25th floor, 32 London Bridge Street
London, SE19SG – United Kingdom
Printed in Croatia

British Library Cataloguing-in-Publication Data
A catalogue record for this book is available from the British Library

Additional hard copies can be obtained from orders@intechopen.com

Hydrogels, Edited by Sajjad Haider and Adnan Haider
p. cm.
Print ISBN 978-1-78923-368-1
Online ISBN 978-1-78923-369-8

For EU product safety concerns:
IN TECH d.o.o., Prolaz Marije Krucifikse Kozulić 3, 51000 Rijeka, Croatia,
info@intechopen.com or visit our website at intechopen.com.

Meet the editors

Dr. Sajjad Haider is an associate professor at the Chemical Engineering Department, King Saud University, Riyadh, Saudi Arabia, since May 2009. He received his MSc degree in 1999; MPhil degree in 2004 from the Institute of Chemical Sciences, University of Peshawar, KPK, Pakistan; and PhD degree in 2009 from the Department of Polymer Science and Engineering, Kyungpook National University, Daegu, South Korea. His research work focuses on the development of carbon nanotubes and biopolymer composites, polymer hydrogel, and electrospun nanofibers for environmental and biomedical applications.

Dr. Adnan Haider is an assistant professor at the Department of Chemistry, Kohat University of Science and Technology. He received his MSc degree in 2010 from the Kohat University of Science and Technology, KPK, Pakistan; PhD degree in 2016 from the Department of Polymer Science and Engineering, Kyungpook National University, Daegu, South Korea; and postdoctoral research in 2017 from the Department of Nano, Medical and Polymer Materials, School of Chemical Engineering, Yeungnam University, South Korea. His research work focuses on the development of scaffolds using polymer hydrogel and electrospun nanofibers for biomedical (tissue regeneration and drug delivery) and environmental applications.

Contents

Preface IX

Section 1 Bio and Synthetic Hydrogels 1

Chapter 1 **Decellularized ECM-Derived Hydrogels: Modification and Properties 3**
Jesús A. Claudio-Rizo, Jorge Delgado, Iraís A. Quintero-Ortega, José L. Mata-Mata and Birzabith Mendoza-Novelo

Chapter 2 **Hydrogels Based on Polyvinylpyrrolidone Copolymers 23**
Oleh Suberlyak and Volodymyr Skorokhoda

Chapter 3 **Superabsorbent 45**
Mohamed Mohamady Ghobashy

Chapter 4 **Hydrogels Applied for Conformance-Improvement Treatment of Oil Reservoirs 69**
Fernanda G. C. Tessarolli, Ailton S. Gomes and Claudia R. E. Mansur

Section 2 Biomedical Application of Hydrogel 89

Chapter 5 **Enhancement of Hydrogels' Properties for Biomedical Applications: Latest Achievements 91**
Ángel Serrano-Aroca

Chapter 6 **Hydrogels Fibers 121**
Javad Foroughi, Azadeh Mirabedini and Holly Warren

Chapter 7 **Obtaining Hydrogels based on PVP/PVAL/Chitosan Containing Pseudoboehmite Nanoparticles for Application in Drugs 141**
Leila Figueiredo de Miranda, Kátia Lucia Gonçalves Cunha, Isabella Tereza Ferro Barbosa, Terezinha Jocelen Masson and Antonio Hortêncio Munhoz Junior

Chapter 8 **Development of PVA/Fe3O4 as Smart Magnetic Hydrogels for Biomedical Applications 159**
Malik Anjelh Baqiya, Ahmad Taufiq, Sunaryono, Munaji, Dita Puspita Sari, Yanurita Dwihapsari and Darminto

Chapter 9 **Hyaluronic-Based Antibacterial Hydrogel Coating for Implantable Biomaterials in Orthopedics and Trauma: From Basic Research to Clinical Applications 179**
Giammona Gaetano, Pitarresi Giuseppe, Palumbo Fabio Salvatore, Maraldi Susanna, Scarponi Sara and Romanò Carlo Luca

Preface

Hydrogels are swollen polymer networks holding water in their structures. By definition, water must constitute at least 10% of the total weight (or volume) for a material to be a hydrogel. However, hydrogels are capable to retain 90% water in their structure. Due to their high water holding capacity, hydrogels have potential applications in processes ranging from industrial to biological fields. Some of the more focused fields include oil reserve treatment, tissue engineering, implants and controlled drug release. Literature on these subjects is expanding. Recently, hydrogels have been explored as potential plugs to control the anisotropic permeability profile of heterogeneous oil reservoirs. Drug delivery systems and tissue engineering implants are the other most studied areas, where hydrogels have been used as 3D materials in dried form. Nonetheless, hydrogels are unique materials and there is always space for more research. A number of original papers, reviews, monographs and technical reports are focusing on the technological aspects of hydrogels to make them more acceptable as a multidisciplinary material. With its distinguished editor and an international team of contributors, the book *Hydrogels* is an outstanding reference for academia, industry and regenerative medicine and environment research scientists.

Dr. Sajjad Haider
Chemical Engineering Department
College of Engineering
King Saud University
Riyadh, Saudi Arabia

Dr. Adnan Haider
Kohat University of Science and Technology
KPK, Kohat, Pakistan

Section 1

Bio and Synthetic Hydrogels

Chapter 1

Decellularized ECM-Derived Hydrogels: Modification and Properties

Jesús A. Claudio-Rizo, Jorge Delgado,
Iraís A. Quintero-Ortega, José L. Mata-Mata and
Birzabith Mendoza-Novelo

Additional information is available at the end of the chapter

http://dx.doi.org/10.5772/intechopen.78331

Abstract

Extracellular matrix (ECM) hydrogels are water-swollen fibrillary three-dimensional (3D) networks where collagen type I is the major component. The hierarchical network formed by the polymerization of tropocollagen molecules with enhanced properties is an attractive template for generating biomaterials. The mammalian tissue source from which collagen is extracted and its consequent modification are variables that impact the physicochemical and biological properties of the collagen network. This chapter has the purpose to provide a review of the research of different strategies to modify and characterize the properties of decellularized ECM-derived hydrogels in the context of safe biomaterials with immunomodulatory properties.

Keywords: ECM, collagen, hydrogel, cross-linking, properties

1. Introduction

Hydrogels are water-swollen polymeric materials with specific three-dimensional (3D) structure. During the last years, hydrogels have been investigated for enhancing biomedical applications. These biomaterials offer a moist environment that can be enriched to provide protection against infections, regulate the inflammation process, promote tissue regeneration, and remove wound exudates [1]. ECM-based hydrogels are promising materials for tissue engineering and regenerative medicine application due to the balance of biochemical and physical characteristics that can be achieved by their modification. Collagen is the main structural protein of the mammalian ECM. It has a favorable impact on blood coagulation, promoting the aggregation of platelets, and the absorption of fluid, and regulating the deposition of other ECM's proteins such as fibrin, laminin, elastin, and fibronectin [2]. Besides, collagen can induce processes of the cell signalization involved in the growth, proliferation, migration, and differentiation of cells. Low inflammatory and cytotoxic responses and high biodegradability are other attractive properties of collagen [3]. The collagen can be extracted from diverse ECMs using acid hydrolysis assisted by proteolytic enzymes. The extracted collagen can be subsequently polymerized under physiological conditions (pH 7, 37°C) to generate a highly hydrated 3D network [4].

The ECM-based hydrogels maintain the biocompatibility and biodegradability associated with the collagen. Diverse authors use to refer ECM hydrogels like collagen hydrogels, as the collagen is the major component inside ECM. However, these biomaterials have poor mechanical properties and fast degradation rate, limiting the range of use in applications such as the loading, encapsulating and controlled delivery of cells or drugs, or as wound care dressings [5]. The structure and mechanics of the ECM hydrogels can be modified by chemical cross-linking (using glutaraldehyde, genipin, carbodiimides, acrylates, oligourethanes, and among others); and/or by physical cross-linking (using freeze-drying cycles, forming interpenetrated networks (IPN) with other proteins or polymers). The selection of the cross-linking strategy has to consider the impact upon the structure-property relationships. After modification, several advances have been reported in the design of ECM-based hydrogels. The delivery of cells and biomolecules, the enhancing of the stiffness, the regulation of the cell-material interactions, the control of the cell fate and function, and the modulation of the environment of both normal and injured/diseased tissues are among them [6]. As shown in **Figure 1**, ECM hydrogels have been studied as substrates for ophthalmology, sponges for burns/wounds, systems for controlled delivery of functional molecules or

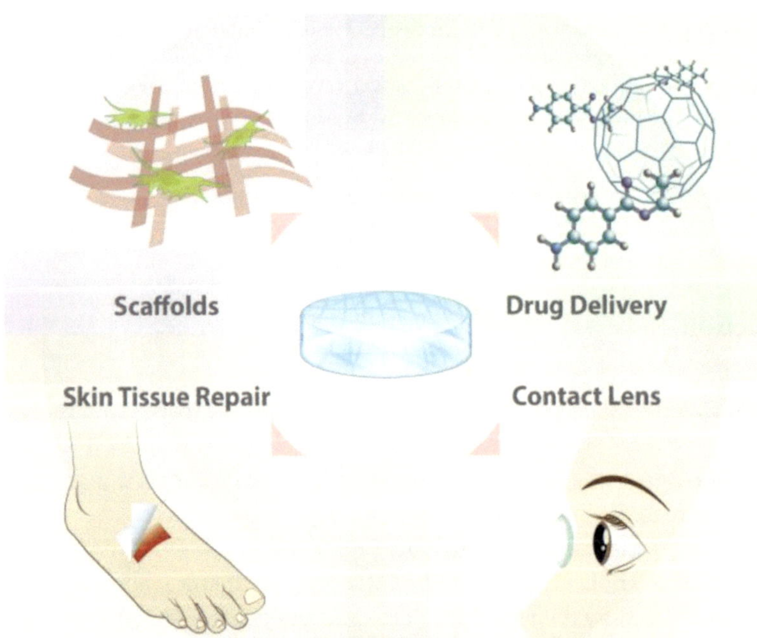

Figure 1. Schematic representation of the biomedical applications of collagen-based hydrogels.

nanoparticles, and matrices for 3D cell culture. They are also investigated for tissue engineering including skin replacement, bone substitutes, and artificial blood vessels and heart valves [7].

In this chapter, a review of the state of the art of different strategies to modify and characterize the properties of natural ECM hydrogels in the context of the biomedical applications is presented. The chapter emphasizes the chemical and physical methods intended to enhance physicochemical properties and immunomodulatory applications of the ECM hydrogels.

2. Methods of preparation of natural ECM-based hydrogels

The versatility of collagen to generate biomedical hydrogels is primarily associated with its complex hierarchical structure originated from its amino acid molecular sequence and the formation of triple helical structures [8]. The collagen polymerization is a self-assembly process of long fibrillar structures, regulated by both electrostatic and hydrophobic interactions that promote the collagen fibers cross-linking [9]. This process is influenced by parameters such as temperature, pH, collagen concentration, and the presence of other biomolecules or polymers [10]. The macroscopic result of the in vitro collagen fibrillogenesis is a 3D water-swollen network.

ECM hydrogels are very suitable materials for biomedical applications due to their good interaction with living tissues, biocompatibility, soft and elastic consistency, high water content, and ECM remaining composition [11]. The swelling in liquid medium gives them the capacity to absorb, retain and release under controlled conditions amounts of water; regulating their structural conformation [12]. The ECM residual composition and the methods of modification of the hydrogels determinate the water uptake capacity and influence their biological and physicochemical properties. This section is dedicated to the discussion of the main characteristics of strategies for the modification of collagen in hydrogel state.

2.1. Importance of the tissue source in the natural ECM-based hydrogels

The ECM is the noncellular component present within all tissues and organs that provides not only essential physical scaffolding for the cellular constituents but also initiates crucial biochemical and biomechanical cues, which are required for tissue morphogenesis, differentiation, and homeostasis [13]. As shown in **Figure 2**, this matrix is composed of a variety of proteins and polysaccharides that are locally secreted and assembled into an organized network in close association with the surface of the cells that produced them [14].

Collagen is the main component of the ECM [15]. The collagen is extracted from diverse ECM by multistep processes including the tissue decellularization and acid hydrolysis assisted with proteases. Among others, collagen has been extracted from porcine dermis [16], bovine pericardium [17], porcine urinary bladder [18], porcine small intestine submucosa [19], bovine Achilles tendon [20], and rat tail tendon [21]. The polymerization of the extracted collagen under physiological conditions (37°C, pH 7) has allowed to develop biomedical hydrogels mimicking the structure and function of the ECM in vitro. The polymerization kinetics and the structural characteristics of the fibrillar collagen gel network are influenced by the residual composition of the ECM. Consequently, the swelling, mechanics, degradation, and biological

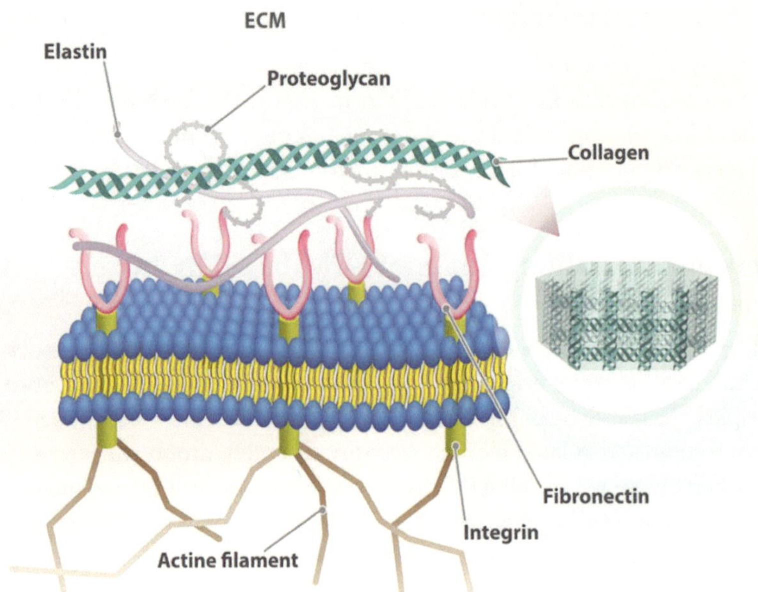

Figure 2. The ECM composition as source for the preparation of collagen-based hydrogels.

response of the ECM hydrogels have shown a direct relationship with the ECM remaining composition [11]. The understanding of the formation process of collagen gels is of relevant importance for the development of strategies capable to synthesize them successfully. The next subsections are focused on those strategies.

2.2. Collagen gel formation in response to change of pH and temperature

The physical methods for the modification of the ECM hydrogels are related to the physical cross-linking of collagen fibers caused by pH, temperature, electrical fields or other physical stimuli, as schematized in **Figure 3** [22]. The advantages of this type of process are the relatively easy manufacture, and the absence of exogenous cross-linking agents, which could reduce the toxicity risks [23]. The variation of pH and temperature of the collagen solution during the in vitro fibrillogenesis produces the collagen cross-linking and increases the fiber size [11, 24]. The temperature-dependent process is reversible [25]. Commonly, the physical methods are not associated with a significant improvement of the mechanical properties of ECM hydrogels, limiting the use of these methods in the preparation of biomedical hydrogels [26].

An interesting physical method to improve the mechanical properties of ECM hydrogels is to apply lyophilization cycles. In this methodology, extracted collagen is incubated at 37°C during 24 h to induce the collagen polymerization, later the hydrogel is frozen at −20°C for 3 h, −80°C for 3 h, and in liquid nitrogen, and then lyophilized. The resulting collagen network demonstrated highly aligned fibrillar features along the scaffold surface, decreased pore size, and increased mechanical properties [27]. However, a major disadvantage related to the

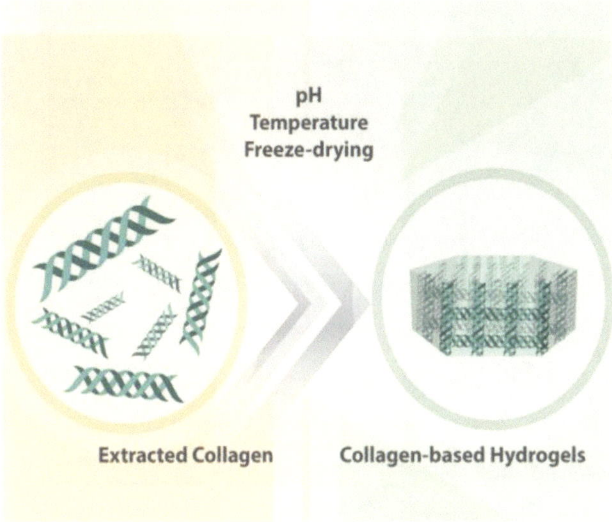

Figure 3. Physical methods for preparation of collagen-based hydrogels.

execution of freeze-drying cycles is the decrease in water uptake of the collagen scaffold [11]. Physical methods do not allow controlling the rate of degradation of collagen-based hydrogels [28]. Therefore, it is still necessary to investigate methods to regulate the characteristics and to expand the use of collagen scaffolds and hydrogels in the medical biotechnology field.

2.3. Interpenetrated networks (IPN) based on collagen and other polymers

IPN hydrogels are based on the physicochemical interactions between the collagen polymeric chains and chains of another type of polymer, as shown in **Figure 4**. The hydrophobic, ionic or hydrogen bonding inside the IPN is responsible for the improved mechanics and degradation behavior. Two examples are the IPNs formed between collagen and chitosan [29], and collagen and polyethylene oxide (PEG) [30]. In these approaches, the ECM extracted collagen is combined with different mass concentrations of polymers, and later this mixture is incubated at 37°C to induce the collagen polymerization. The polymerization process is influenced by the presence of the exogenous polymeric chains altering the collagen fiber size and the physical cross-linking. The IPN hydrogels show poor stability with the change of the temperature and pH [31]; but the enhanced mechanical properties of these biomaterials are adequate for the cell and drug encapsulation [32].

2.4. Chemical cross-linking methods

The search for an ideal procedure to stabilize the structure of collagen maintaining its physical integrity and natural conformation has led to the evaluation of diverse strategies to form covalent bonds. As shown in **Figure 5**, this takes advantage of the conjugation of reactive groups of collagen molecule such as carboxylate (—COO—) and amine (—NH_2) with reactive cross-linkers. Among the most studied processes are the glutaraldehyde (a pentadialdehyde) cross-linking,

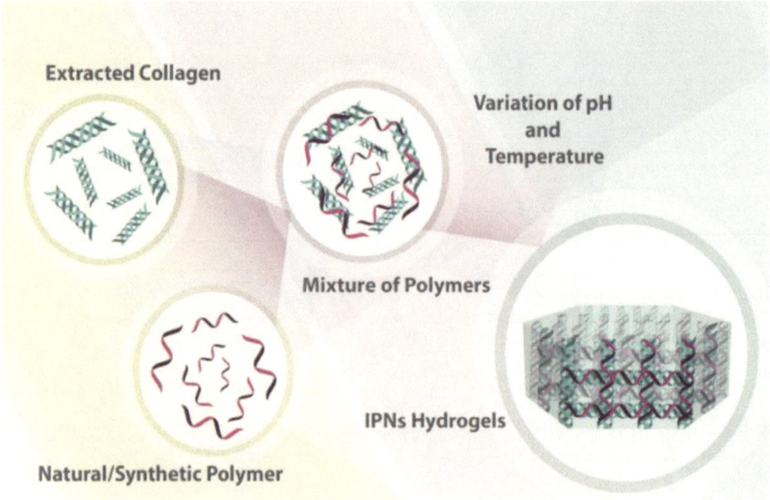

Figure 4. Preparation of hydrogels derived from polymeric IPNs.

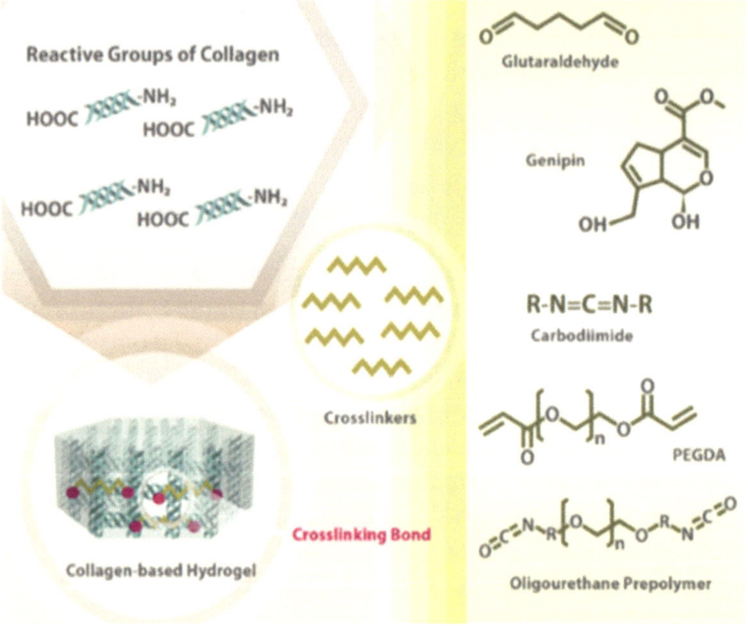

Figure 5. Chemical cross-linking to generate biomedical collagen-based hydrogels.

and the use of carbodiimide 1-Ethyl-3-(3-dimethylaminopropyl)carbodiimide, a water-soluble carbodiimide), genipin, polyethylene glycol diacrylate (PEGDA), and aqueous polyurethane prepolymers. These methods increase the resistance of the hydrogel toward both chemical and enzymatic degradation, reduce its immunogenicity, sterilize and improve its mechanical

Collagen cross-linker	Main characteristic of the process	Advantages	Disadvantages	Ref.
Glutaraldehyde (GA)	The ε-amine groups of collagen yield an imine bond (so-called as Schiff base), when they react with a GA molecule.	The cross-linking reaction is relatively fast; reacting with most ε-amine groups, improving both mechanics and degradation resistance.	Drastic reduction of the biocompatibility.	[29]
1-Ethyl-3-(3-dimethylaminopropyl) carbodiimide (EDAC)	Effective catalyst in the condensation of collagen carboxylic acids with alcohols and amines, without presence of the carbodiimide (so-called zero-length).	The degradation products of these biomaterials do not show cytotoxic character.	The cross-linking reaction is not taken out at physiological conditions.	[33–35]
Genipin	Spontaneous cross-linking by formation of Schiff base is produced. A Michael reaction is involved in this process.	The structure and properties of hydrogels show a direct relationship with the genipin concentration.	Generation of blue residues during the preparation of biomaterials, limiting their transparence and use as 3D culture systems.	[36–38]
Poly(ethylene glycol) diacrylate (PEGDA)	Photo cross-linking based on the formation of covalent linkages among the functional groups acrylamide with the collagen-amines.	Hydrogels show enhanced hydrolytic stability, susceptibility to collagen enzymatic degradation. Mechanical properties depend on time of UV irradiation.	Limitations of use of UV irradiation for applications related to gelation in situ or cell encapsulation.	[39]
Polyurethane prepolymers (Pp)	Pp based on PEG and aliphatic diisocyanates cross-links the collagen chains. The process involves the formation of urea linkages between end-blocked isocyanate of Pp and collagen-amines.	The cross-linking process is taken out at physiological conditions. The structure and properties of collagen hydrogels show a direct relationship with the chemical structure of Pp. Pp accelerates the polymerization.	Higher concentrations of Pp inhibit the collagen polymerization, and decrease its biocompatibility.	[40]

Table 1. Chemical cross-linkers for the preparation of collagen-based hydrogels.

properties. **Table 1** summarizes the main characteristics, advantages and disadvantages of the covalent chemical cross-linking methods.

The elucidation of the impact of the modification upon the structure and properties of the hydrogels derived from decellularized ECM requires the use of a combination of distinct techniques. A forthright correlation between modification and properties is key to balance stability and bioactivity. This chapter thus discusses some aspects of the methods used to discern the characteristics of collagen hydrogels and scaffolds and the implications of their use as safe biomaterials with immunomodulatory properties.

3. Methods for physicochemical characterization of collagen-based hydrogels

A collagen scaffold is a hierarchical, protein-based fibrillary network: after the triple helix formation that conforms tropocollagen, it forms fibrils that align themselves in microfibers and finally, in collagen fibers of a tissue [41]. To correlate the different physicochemical properties observed in a collagen derivative scaffold with its hierarchical network is an attractive challenge partially explored. It is important to notice that the chemical modification of plain tropocollagen is a current practice to tune some properties as the mechanical ones. In general, the characterization methods for the collagen-based materials do not vary when it is chemically modified as we will see in this chapter. On the other hand, for composites that contain collagen, it is intuitive to imagine that the methods to determine their properties could be different depending on the other components of the materials.

3.1. Spectroscopy techniques

Distinct kinds of techniques have been used to characterize the physical and chemical structure of the collagen. Confocal microscopy using the second-harmonic generation [42] and Raman effect [43] have proved to be valuable techniques to determine the presence of the plain collagen in different tissues. The generation of the second harmonic signal in collagen scaffolds is due to the fiber alignment, and it is poor in ECM hydrogels, but it can be enhanced once the collagen is aligned (in natural collagen tissues) or stained. Staining, however, is not recommended because can affect the conformation and interaction among the different components of the scaffold. On the other hand, Raman spectroscopy does not have this limitation and can be used to determine and map collagen in dry and wet samples. In addition, it is sensible to the relative composition of amino acids that confirm the collagen and as a consequence, can be used to determine different kinds of collagen or collagen degradation in time during a disease as cancer for instance [43]. Infrared spectroscopy (IR) can be seen as a complement of the Raman spectroscopy because they are sensitive to the same organic groups. The technical difference is that in Raman, we observe the energy of photons scattered from the sample after excitation using a single wavelength, and in IR, we observe the absorption of photons in a range of wavelengths [44]. In general, IR is used when the sample is not extremely complicated, and signals in the spectra can be assigned to specific interactions in the gel as a chemical modifier as a cross-linker for instance [45, 46]. In such a way, another classical technique that determines chemical interactions among distinct parts of a composite, as ^{1}H-NMR can be used also in ECMs [47].

3.2. Microscopy techniques

Scanning and transmission electronic microscopy (SEM and TEM) are also important techniques for collagen characterization and a first easy access to get the pore size [48], length and width of the fibers as well as shape [49], amount and location of nano- and microstructures of different materials added to the scaffold as can be inorganic salts [50] or nanoparticles [51]. SEM and TEM are excellent characterization techniques for lyophilized scaffolds [52], although in the case of TEM is essential a good handle of the available staining techniques to avoid "artifacts" in the images. However, when those measurements are the reference for wet properties of the scaffold it can be taken only as a guide and other techniques on wet materials are needed to confirm what is being observed.

Using atomic force microscopy (AFM) is possible to observe collagen fibers with different shape and structures on different surfaces [48, 54, 60] as well as to determine micromechanical properties of a collagen hydrogel; both in wet and dry formulations. AFM is microscopy based on the movement of a microtip, in the range of micro- and nanometers, that interacts with the surface at the microscopic level and sense its shape and roughness. The movements of the tip are followed by a laser on it; forming an image of deflection laser intensity. This image, if the deflection force of the tip is known, forms also a map of micromechanical properties.

3.3. X-ray techniques

Another important set of characterization techniques are the X-ray techniques; although their availability depends strongly on the level of development of each particular scientific community in specific countries due to the extensive facilities and economic resources that they need. The more accessible are the X-ray photoelectron spectroscopy (XPS) and the small angle X-ray scattering (SAXS). XPS is able to measure the carbonyl, C═O, and C─C interaction on a collagen surface [53], as well as traces of silicon commonly used now in collagen-based scaffolds [11], for instance. It is in general, the right technique to obtain the gross chemical composition on the collagen surface and a definitive indication for traces of impurities of other elements [54]. Depending on the instrumentation available, wet samples can be measured, since X-rays must work under light vacuum. In addition, it is not expected to see under the surface because the X-ray source is weak and cannot penetrate the sample. On the other hand, scattering X-ray techniques can be used to get the shape, length and width, pore size, and fiber orientation directly from the sample, wet or dry, without further manipulation [55–57]. Depending on the distance to determine, normal X-ray diffraction equipment can also be used [58]. Those techniques are based on the scattering of the X-rays from the sample: a scattering vector is an inverse function of the X-ray wavelength and proportional to the sine of the half the scattering angle. A characteristic distance in the material scatters X-rays of a specific wavelength proportional to the scattering vector and with a different scattering intensity. This intensity is a function of the scattering angle, and it is from where the different properties of the material can be extracted. In general, using smaller scattering angles, it is possible to obtain information about larger distances, as those observed in the collagen when X-rays are used [59].

3.4. Mechanical tests

Mechanical properties (determined at microscopic or macroscopic level) of the collagen are those of a gel or an entangled polymer: basically, oscillatory rheology shows a plateau of the storage modulus (G′) in a frequency ranging between hertz and kilohertz, which can be considered as its Young's modulus, and three times this value can be determined by extensional and compressional experiments of strain versus stress [52]. The convergence of micro- and macromechanical moduli values are not common in the literature [42, 61], although it is expected a similar trend of increment or decrement of the properties toward the same change in a particular variable as can be the cross-linking degree. The Young's modulus can easily vary between cross-linked and uncross-linked collagen one order of magnitude [62, 63]. An important physical parameter directly correlated to the value of the storage modulus is the pore size of the collagen network: the size of the pore is simply the cubic root of the thermal energy (3kT) over the Young's modulus [64]. Alternatively, the pore size distribution of a scaffold can be obtained by analyzing images of thin sections of paraffin-embedded samples obtained by optical microscopy [42] or using electron microscopy as previously explained [48].

Shear flow experiments are useful to obtain the viscosity of the collagen hydrogel precursors, the concentration of the proto-collagen present in a solution, and an estimation of the molecular weight of the minimal structured collagen in solution [65]. It has been also suggested that collagen denaturation can be determined by viscosimetric measurements [66]. Those experiments become important in the case of development of injectable systems because parameters as viscosity [47, 67] and compressibility [52] are important during extrusion. Rheological methods described previously are also convenient to measure the formation of the gel in time: storage (G') (colloquially speaking, how much the viscoelastic material looks like a solid) and loss modulus (G'') (how much the viscoelastic material looks like a liquid) can be determined in an oscillatory rheological measurement to get the gel formation point: where the storage modulus becomes higher than the loss modulus ($G' > G''$) [68].

3.5. Thermal stability test

The denaturation heat and denaturation temperature of a collagen scaffold are obtained from calorimetric experiments commonly using a differential scanning calorimeter (DSC) [67]. Since the technique is based on calorimetric differences sensed by an extremely sensitive electronic device, it is important to consider that minimal differences in the medium concentration (buffer concentration, conductivity of the water used as a solvent, etc.) or during the preparation of the samples (pH, size and shape of the particles, etc.), are observed [48, 69]. Thermal denaturation peak of wet collagen occurs around 50°C, although the heat absorption peak is broad and could start under 20°C before the peak; a straightforward evidence that the collagen has distinct levels of structure. The integral under this endothermal process, that is, energy versus temperature, gives the denaturation heat of the collagen. In general, it has been reported that both denaturation heat and temperature are higher for cross-linked collagen that for uncross-linked collagen [50].

4. Perspectives of the decellularized ECM-based materials in immunomodulation

Biomaterials with immunomodulatory activity are being studied in the context of the repair/regeneration of soft tissues, such as diabetic chronic wounds. Evidences indicate the effect of the characteristics of biomaterials and their (released/biodegraded) by-products over promoting of required immunological responses that could support the wound healing. Moreover, the residual components remaining the animal source as well as the modification of ECM-based materials can elucidate an undesirable response.

4.1. Macrophage polarization in decellularized ECM-based materials

Macrophages are cells of the innate immune system with a dominant effector activity in the injury site after biomaterial implantation. Cross-talk between immune cells activates macrophages after which, they release a variety of signaling molecules. Signaling molecules secreted by macrophages such as cytokines (as interleukins), growth factors as the basic fibroblast growth factor, the vascular endothelial growth factor and the transforming growth factor-beta 1 (bFGF, VEGF and TGF-β1 respectively); and tumor necrosis factor (TNF-α) influence the development of other cell types [70]. In fact, the profile of signaling molecules secretion is commonly evaluated to study the polarization of the macrophage response from an inflammation and tissue injury process to a repair process [71, 72] or to study angiogenesis and scaffold vascularization [73]. Macrophages mediate the healing responses to implanted biomaterials, fundamentally by two outcomes: scar tissue formation (M1M pathway) or regeneration (M2M pathway) [70]. The modulation of the inflammatory response by the physical and chemical properties of biomaterials represents a hypothesis currently assessed in the design of biomaterials intended to the repair/regeneration of soft tissue.

4.2. Impact of the residual composition on the immune response

The goal of the decellularization process of mammalian tissues is to remove its cellular and nuclear components. This aim must be balanced with retention of both the extracellular composition and microstructural characteristics, as much as possible. As result, an incomplete removal of nuclear components has been reported in diverse ECM biomaterials, even in commercial biological implants [74]. The intensity of the host immune response after implantation is heavily influenced by the residual material, which acts as like cell signals [74, 75]. For instance, a decrement in the DNA amount in small intestine submucosa tissue provoked a shift of the M1M proinflammatory macrophage phenotype to the M2M anti-inflammatory one [76]. On the other hand, the tissue regeneration induced by the ECM-based biomaterials has been associated with extracellular residual components such as collagen type I, polysaccharides or basal membrane complex components [77]. Glycosaminoglycans such as hyaluronic acid extracted from brain and urinary bladder have been associated with an up-regulated secretion of anti-inflammatory factors and suppressed secretion of proinflammatory factors, consistent with M2M phenotype macrophages [76]. Moreover, studies revealed that the anionic detergent sodium dodecyl sulfate and nonionic detergent TritonX-100 produce a different impact over the stability of ligands and proteins in the basal membrane complex [80]. The decellularization method and tissue source thus influence the retention of the basal membrane complex components within ECM materials and the bioactivity of them. The bioactivity of ECM-based materials was also evidenced by the differentiation of human monocytes differentiated to macrophages. The higher amounts of interleukin-6 (IL-6), interleukin-8 (IL-8), and monocyte chemoattractant protein-1, but lower amounts of interleukin-10 (IL-10) and interleukin-1 receptor antagonist (IL-1ra) were detected on decellularized pericardium matrix, in comparison with polydimethylsiloxane or polystyrene surfaces [81]. Cellular residual components such as damage-associated molecular patterns (DAMPs, proteins that are retained within the ECM scaffolds) have been considered as bioinductive molecules with a key role in the macrophage polarization [78]. High-mobility group box 1 (a DAMP that functions intracellularly as a DNA binding nuclear protein), detected in ECM biomaterials derived from small intestinal submucosal, and urinary bladder matrix, was correlated with differences in cell proliferation, death, secretion of the immunomodulatory factors [78]. Altogether, reports suggest that decellularization, as the first step in the development of ECM-based hydrogels, and scaffolds, impact the cellular and extracellular components within biomaterials. Consequently, these components become in a key player in the mechanisms of tissue regeneration observed when decellularized ECM materials are used. The ability to support the proliferation and migration of different cells [82], to allow the differentiation of mesenchymal cells [83, 84], and to transit from the inflammatory first steps to a regenerative action [75, 85] are among the mechanisms by which the ECM biomaterials participate. Once animal tissues are decellularized, they are cross-linked to increase their stability, reduce degradation, and immunogenicity. However, the reconstruction of functional tissue appears to be compromised after cross-linking as discussed below.

4.3. Impact of the collagen cross-linking on the immune response

The cross-linking process of ECM-based biomaterials is commonly associated with a detrimental effect on the ultrastructure and composition of the ECM and consequently the biological response [78]. The ability of the decellularized ECM materials to interact with cells is modified by the altered surface chemistry after cross-linking. As discussed above, distinct methods for cross-linking collagen biomaterials have been studied. The understanding and control of cell fate in modified chemically collagen materials is a matter of study. For instance, the cell membrane morphology, cell adhesion and enzymatic activity of the acid phosphatase and esterase of U937 macrophage-like cells have shown to be differentially influenced by the glutaraldehyde cross-linking, and EDAC coupling methods. Glutaraldehyde cross-linking induced an increase in the release of the proinflammatory cytokines IL-1ra, IL-6, IL-10, and TNF-α, unlike to EDAC-cross-linked materials and uncross-linked tissues [86]. Differences in the microenvironment of ECM-based implants cross-linked with glutaraldehyde and diisocyanate (aliphatic) cross-linking methods modified the infiltration of neutrophils and the function of macrophages [87]. A strong proinflammatory milieu was observed in glutaraldehyde-cross-linked materials, while in diisocyanate-cross-linked materials, an anti-inflammatory milieu was seen. The proliferation of immune cell subpopulations was found stronger on both porcine nondecellularized and decellularized materials than on the glutaraldehyde-cross-linked ones [79]. This observation has been reported in the case of cross-linking of tissue-derived heart valves [88]. The integration of this implant type has been associated with a reduced antigenicity by masked immunogenicity [88]. A lack of acute inflammation in dermis-derived implants (fixed with glutaraldehyde at low concentration) both in animal models and humans was observed. Thus, the high concentrations of aldehyde employed in the processing of ECM biomaterials appear to induce a more pronounced and sustained inflammatory response [89]. The dermis-derived implants cross-linked with diisocyanates showed a low chronic inflammatory response after a 20-week period of implantation with both limited collagen degradation and vascular ingrowth [89]. Non–cross-linked ECM materials showed earlier cell infiltration, extracellular matrix (ECM) deposition, scaffold degradation, and neovascularization compared with cross-linked materials, after a 1-month period of implantation. However, after 6 and 12 months, diisocyanate-cross-linked materials showed comparable results compared with the non–cross-linked materials [90]. The cross-linked collagen-derived implants showing an acceptable performance in diverse applications would seem to suggest a degree of tolerance to these materials [90]. However, the tissue remodeling associated with the ECM constituents is yet a challenge to be addressed in the development of new cross-linked ECM biomaterials.

4.4. Immunomodulation with decellularized ECM-based hydrogels

The performance of hernia standard surgical grafts, manufactured from polypropylene, has been improved by the coating of them with decellularized ECM-based hydrogels. This was attributed to the polarization of alternatively-activated and constructive M2Ms macrophages induced by the degradation products from ECM materials, which in turn facilitates migration and myogenesis of skeletal muscle progenitor cells [84]. The migration and proliferation of perivascular stem cells are influenced by the structural components (include a number of partially digested proteoglycans and proteins such as collagens, elastin, laminin, fibronectin, hyaluronan, and heparan) as well as soluble components of hydrogels derived from urinary bladder matrix (include cryptic peptide fragments generated from partial proteolysis of scaffold resident growth factors, and matricellular proteins, e.g., tenascin, osteopontin, and thrombospondin) [76]. The mechanism through the soluble and structural components of ECM-based hydrogels contribute to the host response appears to be different. Both components altered the macrophage behavior but with different fingerprints according to the cytokines secretion profiles [76]. A hernia rodent model study revealed that the implantation of polypropylene meshes coated with ECM hydrogel for a period of 14 days decreased the inflammatory response, which was characterized by the number and distribution of M1Ms around polypropylene fibers, compared to the uncoated devices. After a period of 180 days, the density of mature type I collagen deposited between mesh synthetic fibers was decreased with the coating of ECM hydrogel was used [76]. The coating based on ECM-based hydrogel suggested a low scar tissue deposition on the synthetic mesh, which can be associated with a mitigated chronic inflammatory response, an attenuated M1M response, and an increased M2M/M1M ratio to abdominal defect polypropylene standard grafts [91]. The use of decellularized amniotic membrane tissue combined with poly (urethane-ester) showed a better biocompatibility compared to polypropylene meshes when implanted into abdomen of rabbits over a period of 10 months [92]. Results of in vitro cytocompatibility tests demonstrated that this composite can support primary smooth muscle cells to grow and differentiate, with high proliferation, mitochondrial activity, and special protein expression (α-smooth muscle actin).

5. Final remarks

Mammalian tissues from various sources can be used as biomaterials after modification by decellularization and cross-linking processes. Among these materials, the ECM-based hydrogels seem promising alternatives to modulate the required properties in applications related to biomedicine and tissue engineering. Current approaches usually affect the network structure, physicochemical properties, and biocompatibility of natural ECM-based scaffolds. Thus, a balance between the mechanical and degradation properties and immunology response is a present challenge. In this respect, methodologies based on the combination of the ECM with natural and synthetic polymers, minimizing the removal of the characteristics of the natural ECM, seem to be the best alternatives for this purpose. The structural modification of the natural ECM is related to the variation of its properties; this process can be monitored by a variety of physicochemical techniques, which could provide direct evidence of the structure-property relationship in ECM-based biomaterials. A direct evidence of the ECM properties is definitely a challenge, because some of the most common techniques give only approximations to them. A sample preparation that could include denaturation, drying, staining, etc., can completely change a parameter as the pore size or the fiber alignment. In such a way, new or revised techniques that can be used on undamaged and functional ECM are desirable [93, 94]. Those new techniques where the cellular function is not compromised, will give not only more reliable information about the way ECM interacts in the body, but will open new perspectives on the way to study and prepare ECMs for future applications.

Acknowledgements

Authors thanks funding by the National Council of Science and Technology (CONACyT, México), grant PN_2015-1310.

Conflict of interest

The authors declare no conflict of interest.

Author details

Jesús A. Claudio-Rizo[1], Jorge Delgado[2*], Iraís A. Quintero-Ortega[2], José L. Mata-Mata[2] and Birzabith Mendoza-Novelo[2]

*Address all correspondence to: jorgedel@ugto.mx

1 Polytechnic University of Penjamo, Penjamo, Mexico

2 University of Guanajuato, León, Mexico

References

[1] Varaprasad K et al. A mini review on hydrogels classification and recent developments in miscellaneous applications. Materials Science & Engineering, C: Materials for Biological Applications. 2017;**79**:958-971

[2] Gelse K, Poschl E, Aigner T. Collagens—Structure, function, and biosynthesis. Advanced Drug Delivery Reviews. 2003;**55**(12):1531-1546

[3] Castillo-Briceno P et al. A role for specific collagen motifs during wound healing and inflammatory response of fibroblasts in the teleost fish gilthead seabream. Molecular Immunology. 2011;**48**(6-7):826-834

[4] Saldin LT et al. Extracellular matrix hydrogels from decellularized tissues: Structure and function. Acta Biomaterialia. 2017;**49**:1-15

[5] Tian ZH, Liu WT, Li GY. The microstructure and stability of collagen hydrogel cross-linked by glutaraldehyde. Polymer Degradation and Stability. 2016;**130**:264-270

[6] Claudio-Rizo JA et al. A new method for the preparation of biomedical hydrogels comprised of extracellular matrix and oligourethanes. Biomedical Materials. 2016;**11**:035016

[7] Hinderer S, Layland SL, Schenke-Layland K. ECM and ECM-like materials—Biomaterials for applications in regenerative medicine and cancer therapy. Advanced Drug Delivery Reviews. 2016;**97**:260-269

[8] Cheema U, Ananta M, Mudera Vi. Collagen: Applications of a natural polymer in regenerative medicine. In: Eberli D, editor. Regenerative Medicine and Tissue Engineering—Cells and Biomaterials. InTech; 2011. pp. 287-297

[9] Theocharis AD et al. Extracellular matrix structure. Advanced Drug Delivery Reviews. 2016;**97**:4-27

[10] Ghazanfari S, Khademhosseini A, Smit TH. Mechanisms of lamellar collagen formation in connective tissues. Biomaterials. 2016;**97**:74-84

[11] Claudio-Rizo JA et al. Influence of residual composition on the structure and properties of extracellular matrix derived hydrogels. Materials Science & Engineering, C: Materials for Biological Applications. 2017;**79**:793-801

[12] Annabi N et al. Controlling the porosity and microarchitecture of hydrogels for tissue engineering. Tissue Engineering Part B-Reviews. 2010;**16**(4):371-383

[13] Kular JK, Basu S, Sharma RI. The extracellular matrix: Structure, composition, age-related differences, tools for analysis and applications for tissue engineering. Journal of Tissue Engineering. 2014;**5**:2041731414557112

[14] Morris AH, Kyriakides TR. Matricellular proteins and biomaterials. Matrix Biology. 2014; **37**:183-191

[15] Eyre DR. Collagen—Molecular diversity in the body's protein scaffold. Science. 1980;**207**(4437):1315-1322

[16] Wolf MT et al. A hydrogel derived from decellularized dermal extracellular matrix. Biomaterials. 2012;**33**(29):7028-7038

[17] Climov M et al. Chapter 8—Natural Biomaterials for Skin Tissue Engineering A2—Albanna, Mohammad Z. Boston: Academic Press; 2016. pp. 145-161

[18] Freytes DO et al. Preparation and rheological characterization of a gel form of the porcine urinary bladder matrix. Biomaterials. 2008;**29**(11):1630-1637

[19] Lee C et al. Human umbilical cord blood-derived mesenchymal stromal cells and small intestinal submucosa hydrogel composite promotes combined radiation-wound healing of mice. Cytotherapy. 2017;**19**(9):1048-1059

[20] Banerjee P, Mehta A, Shanthi C. Investigation into the cyto-protective and wound healing properties of cryptic peptides from bovine Achilles tendon collagen. Chemico-Biological Interactions. 2014;**211**:1-10

[21] Bornstein MB. Reconstituted rattail collagen used as substrate for tissue cultures on coverslips in Maximow slides and roller tubes. Laboratory Investigation. 1958;**7**(2):134-137

[22] Harris JR, Soliakov A, Lewis RJ. In vitro fibrillogenesis of collagen type I in varying ionic and pH conditions. Micron. 2013;**49**:60-68

[23] Yunoki S, Matsuda T. Simultaneous processing of fibril formation and cross-linking improves mechanical properties of collagen. Biomacromolecules. 2008;**9**(3):879-885

[24] Lv Q et al. Fibroin/collagen hybrid hydrogels with crosslinking method: Preparation, properties, and cytocompatibility. Journal of Biomedical Materials Research Part A. 2008;**84A**(1):198-207

[25] Perez CMR, Rank LA, Chmielewski J. Tuning the thermosensitive properties of hybrid collagen peptide-polymer hydrogels. Chemical Communications. 2014;**50**(60):8174-8176

[26] Walters BD, Stegemann JP. Strategies for directing the structure and function of three-dimensional collagen biomaterials across length scales. Acta Biomaterialia. 2014;**10**(4): 1488-1501

[27] Wang L, Stegemann JP. Thermogelling chitosan and collagen composite hydrogels initiated with beta-glycerophosphate for bone tissue engineering. Biomaterials. 2010;**31**(14): 3976-3985

[28] Thambi T, Li Y, Lee DS. Injectable hydrogels for sustained release of therapeutic agents. 2017;**267**:57-66

[29] Wu X et al. Preparation and assessment of glutaraldehyde-crosslinked collagen-chitosan hydrogels for adipose tissue engineering. Journal of Biomedical Materials Research Part A. 2007;**81A**(1):59-65

[30] Bartlett RS, Thibeault SL, Prestwich GD. Therapeutic potential of gel-based injectables for vocal fold regeneration. Biomedical Materials. 2012;**7**(2):024103

[31] Sharabi M et al. A new class of bio-composite materials of unique collagen fibers. Journal of the Mechanical Behavior of Biomedical Materials. 2014;**36**:71-81

[32] Matricardi P et al. Interpenetrating polymer networks polysaccharide hydrogels for drug delivery and tissue engineering. Advanced Drug Delivery Reviews. 2013;**65**(9):1172-1187

[33] Sheehan JC, Hlavka JJ. The use of water-soluble and basic carbodiimides in peptide synthesis. The Journal of Organic Chemistry. 1956;**21**(4):439-441

[34] Sheehan JC, Hlavka JJ. The cross-linking of gelatin using a water-soluble carbodiimide. Journal of the American Chemical Society. 1957;**79**(16):4528-4529

[35] Rafat M et al. PEG-stabilized carbodiimide crosslinked collagen-chitosan hydrogels for corneal tissue engineering. Biomaterials. 2008;**29**(29):3960-3972

[36] Yoo JS et al. Study on genipin: A new alternative natural crosslinking agent for fixing heterograft tissue. The Korean Journal of Thoracic and Cardiovascular Surgery. 2011;**44**(3):197-207

[37] Grolik M et al. Hydrogel membranes based on genipin-cross-linked chitosan blends for corneal epithelium tissue engineering. Journal of Materials Science. Materials in Medicine. 2012;**23**(8):1991-2000

[38] Yunoki S, Ohyabu Y, Hatayama H. Temperature-responsive gelation of type I collagen solutions involving fibril formation and genipin crosslinking as a potential injectable hydrogel. International Journal of Biomaterials. 2013;**2013**:14

[39] Singh RK, Seliktar D, Putnam AJ. Capillary morphogenesis in PEG-collagen hydrogels. Biomaterials. 2013;**34**(37):9331-9340

[40] Mendoza-Novelo B et al. Synthesis and characterization of protected oligourethanes as crosslinkers of collagen-based scaffolds. Journal of Materials Chemistry B. 2014;**2**(19): 2874-2882

[41] Buehler MJ. Nature designs tough collagen: Explaining the nanostructure of collagen fibrils. Proceedings of the National Academy of Sciences of the United States of America. 2006;**103**(33):12285-12290

[42] Yannas IV et al. Biologically active collagen-based scaffolds: Advances in processing and characterization. Philosophical Transactions of the Royal Society A-Mathematical Physical and Engineering Sciences. 2010;**368**(1917):2123-2139

[43] Nguyen TT et al. Characterization of type I and IV collagens by Raman microspectroscopy: Identification of spectral markers of the dermo-epidermal junction. Spectroscopy. 2012;**27**(5-6):421-427

[44] Confocal Raman Microscopy. Springer Series in Optical Sciences. Vol. 158. Springer-Verlag Berlin Heidelberg; 2011

[45] Pamfil D, Schick C, Vasile C. New hydrogels based on substituted anhydride modified collagen and 2-hydroxyethyl methacrylate. Synthesis and characterization. Industrial & Engineering Chemistry Research. 2014;**53**(28):11239-11248

[46] Su XR et al. Characterization of acid-soluble collagen from the coelomic wall of Sipunculida. Food Hydrocolloids. 2009;**23**(8):2190-2194

[47] Leyva-Gomez G et al. Physicochemical and functional characterization of the collagen-polyvinylpyrrolidone copolymer. Journal of Physical Chemistry B. 2014;**118**(31):9272-9283

[48] Li CH et al. Characterization of acylated pepsin-solubilized collagen with better surface activity. International Journal of Biological Macromolecules. 2013;**57**:92-98

[49] Bet MR, Goissis G, Lacerda CA. Characterization of polyanionic collagen prepared by selective hydrolysis of asparagine and glutamine carboxyamide side chains. Biomacromolecules. 2001;**2**(4):1074-1079

[50] Claudio-Rizo JA et al. Improved properties of composite collagen hydrogels: Protected oligourethanes and silica particles as modulators. Journal of Materials Chemistry B. 2016;**4**(40):6497-6509

[51] Nakano A et al. Preparation and characterization of complex gel of type I collagen and aluminosilicate containing imogolite nanofibers. Journal of Applied Polymer Science. 2010;**118**(4):2284-2290

[52] Zhao LL et al. A novel smart injectable hydrogel prepared by microbial transglutaminase and human-like collagen: Its characterization and biocompatibility. Materials Science & Engineering, C: Materials for Biological Applications. 2016;**68**:317-326

[53] Cote MF et al. Denatured collagen as support for a FGF-2 delivery system: Physicochemical characterizations and in vitro release kinetics and bioactivity. Biomaterials. 2004;**25**(17): 3761-3772

[54] Adamczak M et al. Surface characterization, collagen adsorption and cell behaviour on poly(L-lactide-co-glycolide). Acta of Bioengineering and Biomechanics. 2011;**13**(3):63-75

[55] Wells HC et al. Collagen fibril structure and strength in acellular dermal matrix materials of bovine, porcine, and human origin. ACS Biomaterials Science & Engineering. 2015;**1**(10):1026-1038

[56] Rubina MS et al. Collagen-chitosan scaffold modified with Au and Ag nanoparticles: Synthesis and structure. Applied Surface Science. 2016;**366**:365-371

[57] Hanazaki Y et al. Multiscale analysis of changes in an anisotropic collagen gel structure by culturing osteoblasts. ACS Applied Materials & Interfaces. 2013;**5**(13):5937-5946

[58] Fauzi MB et al. Ovine tendon collagen: Extraction, characterisation and fabrication of thin films for tissue engineering applications. Materials Science & Engineering, C: Materials for Biological Applications. 2016;**68**:163-171

[59] Fitter J, Gutberlet T, Katsaras J. Neutron Scattering in Biology. Biological and Medical Physics, Biomedical Engineering. Springer-Verlag Berlin Heidelberg; 2006

[60] Dufrene YF, Marchal TG, Rouxhet PG. Influence of substratum surface properties on the organization of adsorbed collagen films: In situ characterization by atomic force microscopy. Langmuir. 1999;**15**(8):2871-2878

[61] Di Benedetto C et al. Production, characterization and biocompatibility of marine collagen matrices from an alternative and sustainable source: The sea urchin *Paracentrotus lividus*. Marine Drugs. 2014;**12**(9):4912-4933

[62] O Halloran DM et al. Characterization of a microbial transglutaminase cross-linked type II collagen scaffold. Tissue Engineering. 2006;**12**(6):1467-1474

[63] Garcia Y et al. In vitro characterization of a collagen scaffold enzymatically cross-linked with a tailored elastin-like polymer. Tissue Engineering Part A. 2009;**15**(4):887-899

[64] Macosko CW. Rheology: Principles, Measurements and Applications. USA; 1994

[65] Sulea D et al. Characterization and in vitro release of chlorhexidine digluconate comprised in type i collagen hydrogels. Revue Roumaine de Chimie. 2010;**55**(9):543-551

[66] Rama S, Chandrakasan G. Physicochemical characterization and molecular-organization of the collagen from the skin of an air-breathing fish (*Ophiocephalus-striatus*). Journal of Biosciences. 1983;**5**(2):147-154

[67] Huang CY et al. Isolation and characterization of fish scale collagen from tilapia (*Oreochromis* sp.) by a novel extrusion-hydro-extraction process. Food Chemistry. 2016;**190**:997-1006

[68] Yang YL, Kaufman LJ. Rheology and confocal reflectance microscopy as probes of mechanical properties and structure during collagen and collagen/hyaluronan self-assembly. Biophysical Journal. 2009;**96**(4):1566-1585

[69] Wolf KL, Sobral PJA, Telis VRN. Physicochemical characterization of collagen fibers and collagen powder for self-composite film production. Food Hydrocolloids. 2009;**23**(7): 1886-1894

[70] Franz S et al. Immune responses to implants—A review of the implications for the design of immunomodulatory biomaterials. Biomaterials. 2011;**32**(28):6692-6709

[71] Jaguin M et al. Polarization profiles of human M-CSF-generated macrophages and comparison of M1-markers in classically activated macrophages from GM-CSF and M-CSF origin. Cellular Immunology. 2013;**281**(1):51-61

[72] Spiller KL et al. The role of macrophage phenotype in vascularization of tissue engineering scaffolds. Biomaterials. 2014;**35**(15):4477-4488

[73] Dohle E et al. Macrophage-mediated angiogenic activation of outgrowth endothelial cells in co-culture with primary osteoblasts. European Cells & Materials. 2014;**27**:149-165

[74] Brown BN et al. Macrophage phenotype and remodeling outcomes in response to biologic scaffolds with and without a cellular component. Biomaterials. 2009;**30**(8):1482-1491

[75] Sicari BM et al. The promotion of a constructive macrophage phenotype by solubilized extracellular matrix. Biomaterials. 2014;**35**(30):8605-8612

[76] Slivka PF et al. Fractionation of an ECM hydrogel into structural and soluble components reveals distinctive roles in regulating macrophage behavior. Biomaterials Science. 2014;**2**(10):1521-1534

[77] Sanchez-Sanchez R et al. Generation of two biological wound dressings as a potential delivery system of human adipose-derived mesenchymal stem cells. ASAIO Journal. 2015;**61**(6):718-725

[78] Daly KA et al. Damage associated molecular patterns within xenogeneic biologic scaffolds and their effects on host remodeling. Biomaterials. 2012;**33**(1):91-101

[79] Bayrak A et al. Human immune responses to porcine xenogeneic matrices and their extracellular matrix constituents in vitro. Biomaterials. 2010;**31**(14):3793-3803

[80] Faulk DM et al. The effect of detergents on the basement membrane complex of a biologic scaffold material. Acta Biomaterialia. 2014;**10**(1):183-193

[81] Ariganello MB et al. Macrophage differentiation and polarization on a decellularized pericardial biomaterial. Biomaterials. 2011;**32**(2):439-449

[82] Eitan Y et al. Acellular cardiac extracellular matrix as a scaffold for tissue engineering: In vitro cell support, remodeling, and biocompatibility. Tissue Engineering Part C-Methods. 2010;**16**(4):671-683

[83] Rajabi-Zeleti S et al. The behavior of cardiac progenitor cells on macroporous pericardium-derived scaffolds. Biomaterials. 2014;**35**(3):970-982

[84] Lu T-Y et al. Repopulation of decellularized mouse heart with human induced pluripotent stem cell-derived cardiovascular progenitor cells. Nature Communications. 2013;**4**:2307

[85] Brown BN, Sicari BM, Badylak SE. Rethinking regenerative medicine: A macropnage-centered approach. Frontiers in Immunology. 2014;**5**:510

[86] McDade JK et al. Interactions of U937 macrophage-like cells with decellularized pericardial matrix materials: Influence of crosslinking treatment. Acta Biomaterialia. 2013;**9**(7):7191-7199

[87] Ye Q et al. The relationship between collagen scaffold cross-linking agents and neutrophils in the foreign body reaction. Biomaterials. 2010;**31**(35):9192-9201

[88] Zilla P et al. Prosthetic heart valves: Catering for the few. Biomaterials. 2008;**29**(4):385-406

[89] Macleod TM et al. Histological evaluation of Permacol™ as a subcutaneous implant over a 20-week period in the rat model. British Journal of Plastic Surgery. 2005;**58**(4):518-532

[90] Deeken CR et al. Histologic and biomechanical evaluation of crosslinked and non-crosslinked biologic meshes in a porcine model of ventral incisional hernia repair. Journal of the American College of Surgeons. 2011;**212**(5):880-888

[91] Wolf MT et al. Macrophage polarization in response to ECM coated polypropylene mesh. Biomaterials. 2014;**35**(25):6838-6849

[92] Shi P et al. Biocompatible surgical meshes based on decellularized human amniotic membrane. Materials Science and Engineering: C. 2015;**54**:112-119

[93] Casavant BP et al. Suspended microfluidics. Proceedings of the National Academy of Sciences of the United States of America. 2013;**110**(25):10111-10116

[94] Koster S et al. Visualization of flow-aligned type I collagen self-assembly in tunable pH gradients. Langmuir. 2007;**23**(2):357-359

Hydrogels Based on Polyvinylpyrrolidone Copolymers

Oleh Suberlyak and Volodymyr Skorokhoda

Additional information is available at the end of the chapter

http://dx.doi.org/10.5772/intechopen.72082

Abstract

The role of polyvinylpyrrolidone (PVP) complex formation with water-soluble 2-hydroxyalkyl methacrylates is described. The impact of the complexation on both the polymerization kinetics and the formation of a copolymer structure initiated by radical initiators has been studied. The activating effect of iron(II) and iron(III) sulfates has been revealed for the initiator-free polymerization of the formulation. An analytical approach to determining the molecular weight of the chain fragments located between two neighboring cross-linking nodes in the polymer network (Mn) has been developed depending on the values of the stability constant (Kst) for the charge-transfer complexes. The basic regularities of hydrogels obtaining based on PVP copolymers with high sorption capacity and diffusion characteristics are presented. The main directions of practical application of synthesized hydrogels are considered.

Keywords: hydrogels, polyvinylpyrrolidone, complex with charge transfer, cross-linked copolymers, permeability membranes, capsulation particles, drugs, soft contact lenses, biomedical, properties

1. Introduction

The concentration of colloid polymer solutions is accompanied by increasing viscosity up to a critical value when a gel is formed. A gel (a jellylike material) is a system which exhibits no flow and is based on a fluctuation polymer network swollen in a solvent. The formation of gels is accompanied by the appearance of physical nodes between macromolecular chains. The stability of fluctuation nodes and, therefore, the gel stability increase with increasing energy of the interaction between solvent molecules and polymer chains. In the case of using aqueous solutions of natural or synthetic polymers, a "hydrogel" is formed. This is a hydrogel of the "second type." Such a product consists of two phases and is unstable. During significant change of temperature or dynamic load it divides into two phases that hydrogel—formed a syneresis process occurs [1]. That is why a gel of the "second type" cannot be recommended for long-lasting exploitation under variable conditions.

Polymeric gels of the "first type" are formed upon swelling of chemically cross-linked hydrogels, and their matrix consists of macromolecule segments located between chemical cross-linking nodes. It leads to the formation of chemically cross-linked network that swells due to the sorption of a solvent. Chemical bonds between macromolecules provide non-fluidity of the system. Network swells partially as a result of change of the segment conformation under the effect of a solvent.

Conformational static macromolecular coil (**Figure 1**) of chain segments between cross-linking nodes causes significant reversible deformation which corresponds to highly elastic deformation under the influence of external force field. Hydrogels are formed during the swelling of chemically cross-linked highly hydrophilic polymers in water. A large number of hydrogels, obtained through the polymerization in water or in bulk with the following swelling of the synthesized polymer in water, are known [2].

Water-soluble monomers such as vinylpyrrolidone [3], hydroxyalkyl (met)acrylates (HAMA) [4–6] and their homologs (C_3–C_{13}) [7], propylene glycol methacrylates [8], etc. are used for the synthesis of a polymer matrix.

In the method [9], a chemically cross-linked structure is formed due to the usage of a bifunctional monomer of similar nature in reaction mass. Cross-linking agents (CA), which are used for the polymerization of monofunctional monomers, are bis-(met)acrylates of glycols [10–14], bis-allylic esters [15, 16], triallyl cyanurate [17], dialdehydes [18], and polyethylene glycol dimethacrylates [19].

The number of CA affects the degree of polymer matrix cross-linking and molecular weight of intermolecular crosslinks [20].

Content of water can vary from 5 to 90% depending on the quantity of cross-linking agent and its molecular weight [20, 21]. The quantity of CA with low molecular weight, such as dimethacrylates, can be 0.25–2%, which would provide a sufficient amount of water in a hydrogel [22–24]. It has been mentioned that hydrogels based on hydroxyalkyl (met)acrylates, used for production of contact lenses, have quite low oxygen permeability. Oxygen permeability of hydrogel with 28% of water is $35 \cdot 10^{-10}$ $cm^2 \cdot mL\ O_2/mL \cdot s \cdot mm$.

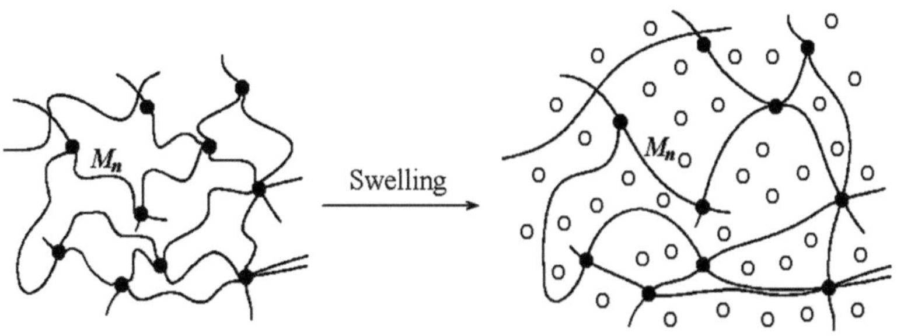

Figure 1. Schematic diagram of swelling of hydrogel [(•) cross-linking node and (○) water molecule].

It has been stated [25] that oxygen permeability depends just on the water content and does not depend on the chemical structure of a hydrogel matrix.

A highly hydrophilic matrix of a hydrogel can also be obtained due to chemical cross-linking of water-soluble polymers. For example, polyvinyl alcohol (PVA) cross-linked by heating in the presence of sodium tetraborate [26] or by initiated graft polymerization, in particular, PVA with glycidyl methacrylate [27]. To this end, polyvinylpyridine, poly(ethylene glycol), and hydroxypropyl cellulose are also applied besides PVA. [13, 15, 28–31]. Such polymers are mainly used for the reduction of internal tensions due to them washing out during hydration process and increase of matrix-free volume that decreases spatial obstacles for the conformational changes of structured polymer chains.

Method of grafted copolymerization of water-soluble monomers on polyvinylpyrrolidone (PVP) appears to be particularly promising with significant possibilities of hydrogel polymeric matrix formation [32, 33]. PVP is used by itself as a sorbent, a thickener of cosmetic ointments and for encapsulation of medical drugs [34]. Due to its high surface energy, PVP is also an attractive (a promising) substance in the formation of metal nanopowders [35, 36] as well as silicate nanopowders from corresponding solutions [37]. PVP keeps adsorbed drugs on the pyrrolidone rings of the macromolecule [38, 39].

Macromolecule of PVP in a free state has a helicoidal structure with pyrrolidone rings outside, which promotes the interaction of the peptide groups with substances by complex formation. PVP is characterized by high complexation ability. It forms complexes with organic and inorganic electron donor as well as electron acceptor compounds. Complexes form highly polarized peptide groups of the pyrrolidone rings due to mesomeric effect. Specific role is played by complexes of PVP with vinyl monomers which polymerize in the presence of PVP.

2. Role of polyvinylpyrrolidone in the kinetics and formation of structure

Important and notable property of PVP is the ability to form complexes [40–42]. This ability significantly influences the kinetic of polymerization and the formation of a polymeric matrix structure in the process of hydrogel synthesis.

As it has been shown in the research papers [43, 44], PVP can form charge-transfer complexes not only with medical drugs but also with water-soluble vinyl monomers. The results of spectral analysis and quantum mechanical calculations with the application of package Chem3D [45] shows that –C = C– bond of a monomer molecule, negative charge of which significantly changes, and nitrogen atom of the pyrrolidone cycle (–N–), the charge of which increases from +0.35 to +0.42, both participate in the formation of a complex (**Figure 2**).

The complex was characterized by the constant of complex stability (K_{st}), and its value increases with the presence of water or primary alcohol groups [43, 46]. Based on this information, the structure of a charge-transfer complex (CTC) with, for example, 2-hydroxyethyl methacrylate (HEMA) was substantiated [33, 46].

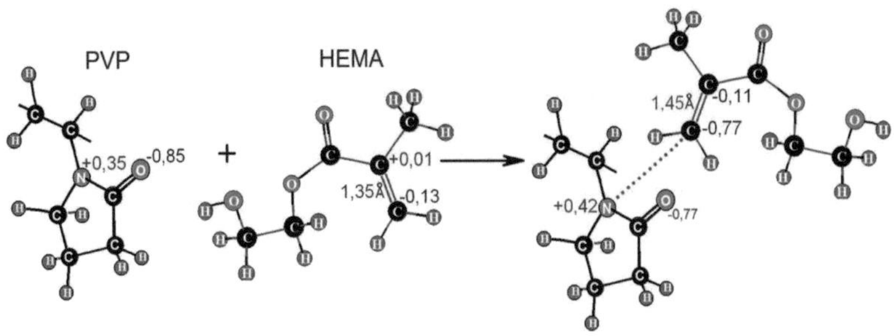

Figure 2. Quantum mechanical model of interaction by PVP with HEMA.

The constant of stability of CTC represents fraction of quantity of the molecules of a reaction mixture (molecules of monomer and elementary links of PVP), which form CTC, to their general quantity in the volume. The change of optical density of diluted solutions of the monomer and PVP in a chosen solvent is determined.

As a result of the monomer molecule solvation on the PVP macrochains through CTC, the rate of HEMA [43] polymerization increases significantly. The rate constant of the polymerization significantly depends on K_{st} of CTC. The polymerization rate increases with the increase of K_{st} of CTC with the maximum at the equimolar ratio of a proton donor (H_2O) and segments of PVP (**Table 1**) [47].

An activation effect of PVP can be observed in proton donor solvents and allows the polymerization without initiators of radical type [48].

The results prove the matrix mechanism of polymerization—local concentration of monomer molecules activated with CTC on the chains of PVP.

Solvent	K_{st} (dm³/mole)	Extinction coefficient (dm³/(mole·cm))	Viscosity[1], ($\eta \cdot 10^3$, Pa·sec)	$V^1 \cdot 10^4$ (mole/(dm³·sec) (at 60°C))	Degree of PVP graft, P, %
Dimethyl sulfoxide	0	–	2.4	0	–
Cyclohexanol	0.06	20.8	17.6	0.6	11
Butanol	0.12	10.0	2.1	0.8	–
Ethylene glycol	0.17	5.6	14.4	1.1	14
Diethylene glycol	0.21	2.1	22.3	1.5	15
Water	0.28	5.3	5.3	3.8	18

[1] Comment: HEMA-PVP: solvent = 9:1:10 mass parts (without initiator), initiator-benzoperoxide.

Table 1. Influence of the nature of solvents on the stability constant of the complex and on the polymerization rate (V) of HEMA-PVP composition.

Such mechanism allows to explain the formation of grafted and few structured PVP copolymer [62]:

1. Adsorption of an initiator and solvation of a monomer on PVP macromolecules and formation of change transfer complex
2. Initiation
3. Chain growth
4. The chain transfer to the PVP as from initial radical $R^{\bullet}$ and from macroradical $R_m^{\bullet}$
5. Graft copolymerization on PVP

Obtaining of grafted copolymers as a result of macroradical combination.

The degree of grafting of PVP depends on the nature of a complex-forming solvent and the nature of an initiator of polymerization of HEMA-PVP compositions (**Table 1**).

Matrix effect increases with the increase of hydroxyl group number in the solution and with the increase of the molecular weight of a proton donor. As a result, CTC with polyvinyl alcohol as a proton donor was found to have significant activating ability [49].

This method allowed to obtain hydrogels with higher mechanical resistance based on the combined matrix PVP:PVA [49, 50]. Complex based on the PVA and PVP shows the highest efficiency at the ratio of 2:1 (**Table 2**).

Efficiency of grafting (f) (inclusion of PVP macromolecules into the copolymer structure or its chemical cross-linking), and also cross-linking degree of macrochains in a polymer network (M_n), first of all depends on the value of stability constant of CTC (**Figure 3**) [47]. During polymerization of HEMA:PVP composition, K_{st} was changed by the replacement of certain amount of water for the acceptor of protons (dimethyl sulfoxide (DMSO)).

Composition of forming solution (mass parts)				M_n (kg/mole)	Tensile strength (σ, MPa)	Relative elongation at rupture (ε, %)
PVP	PVA	HEMA	H_2O			
–	30	30	200	15	10.5	500
25	5	30	200	175	0.29	281
25	10	30	200	180	0.45	390
30	15	15	200	185	0.21	260
25	10	30	200 DMSO	195	0.49	295
25	10	15 + 15 GMA	100 + 100 DMSO	146	0.87	510

Comments: M_n, molecular weight mass of the fragment of macrochain between two neighboring cross-linking nodes; GMA, glycidyl methacrylate.

Table 2. Physico-mechanical properties of membranes based on the hydrogels in hydrated state.

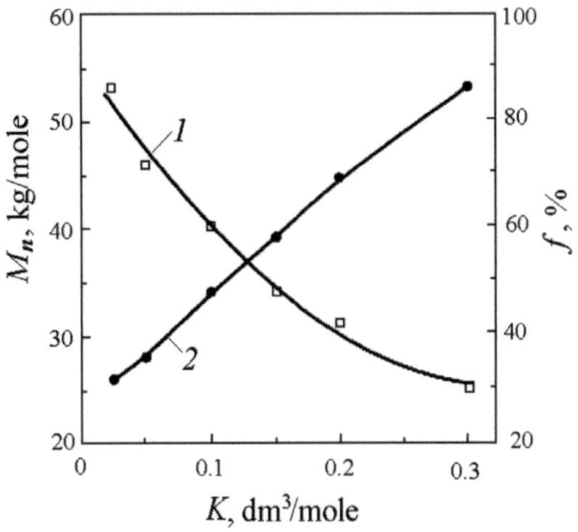

Figure 3. Dependences of the internodal molecular mass M_n (1) and grafting efficiency f (2) on the constant K of complex formation between 2-hydroxyethyl methacrylate and polyvinylpyrrolidone.

The molecular weight of the fragment of macrochain between two neighboring cross-linking nodes has been calculated according to the following formula:

$$M_n = \frac{L^5 \rho_p v_s}{0.5 - \mu} \tag{1}$$

where L is the linear swelling coefficient, ρ_p is the polymer density (kg/m³), v_s is the molar volume of the solvent (m³/(kg·mole)), and μ is the parameter of polymer-liquid interaction:

$$\mu = 0.5 - \frac{v_s \sigma_\infty L^4}{RT(\lambda^2 - \lambda^{-1})'} \tag{2}$$

where σ_∞ is the equilibrium voltage (kgf/m²):

$$\lambda = 1 + \varepsilon, 0 < \varepsilon < 0.3 \tag{3}$$

where ε is the equilibrium voltage strain.

Profitability for practical realization under the circumstances of predicted synthesis dependence of the M_n on the amount of DMSO as proton acceptor to water (A)

$$A = \ln\frac{6.25}{K_{st}} - 6.875 \tag{4}$$

Using this dependence, the exponential dependence of M_n on K_{st} is offered:

$$M_n = M_n^0 \cdot \exp(-2.9\, K_{st}) \tag{5}$$

where M_n^0 is the molecular weight mass of the fragment of macrochain between two neighboring cross-linking nodes by $K_{st} = 0$ (in DMSO).

Contents of the components, mass parts				M_n (kg/mole)	K_{lt} (%)	σ (MPa)	ε (%)	$k \cdot 10^4$ ($m^3/(m^2 \cdot h)$)
HAMA[1]	PVP	H_2O	DMSO					
80	20	100	0	24	95	0.40	235	52
80	20	99	1	25	95	0.40	235	53
80	20	90	10	31	95	0.41	240	57
80	20	80	20	41	95	0.41	245	63
80	20	70	30	52	95	0.42	250	70
80	20	60	40	65	95	0.42	255	76
80	20	40	60	67	—	—	—	77
80	20	80	20	38	96	0.43	230	57
80	20	80	20	44	94	0.40	240	70

Comments: In examples 1–7, we used 2-hydroxyethyl methacrylate as HAMA; in example 8, we used 2-hydroxypropyl methacrylate; and in example 9, we used 2-hydroxypropyl acrylate. DMSO, dimethyl sulfoxide; K_{lt} is the luminous transmission factor, and k is the permeability coefficient of water.

Table 3. Influence of the amount and nature of solvents on the properties of hydrogels.

The dependencies which are appropriate for analytical forecast of the copolymer structure have been proposed. The experimental results of synthesis of hydrogels based on HEMA/PVP at the various amounts of DMSO in the initial composition have been obtained (**Table 3**).

3. Effect of the amount of grafted PVP on the sorption parameters of copolymers

Hydrogels based on the structured hydrophilic copolymers can be obtained due to water sorption. Water sorption by this (co)polymers occurs up to equilibrium-limited swelling of polymeric matrix due to the presence of hydrophilic groups –OH, –C = O, –NH–, and $-NH_2$ in their structure. This process is going with different rates depending on the hydrophilic properties of polymer network and volume (bulk) of block sample.

Equilibrium swelling is characterized by the coefficient of swelling:

Volume Mass Linear

$$K_V = \frac{V_K}{V_0} \quad K_M = \frac{m_{max}}{m_0} \quad K_L = \frac{L_{max}}{L_0} \tag{6}$$

The coefficient of linear swelling is within 1.13…1.20 [51], and the amount of water content is within 20…90%, which can be calculated with the equation:

$$W_{H_2O} = \frac{m_{max} - m_0}{m_{max}} \cdot 100\% \tag{7}$$

where m_{max} is the mass of the sample after swelling and m_0 is the initial mass of the sample before swelling.

Hydrogel is also characterized by water sorption—the amount of water that can be sorbed by dry sample during swelling up to reaching the equilibrium state:

$$W_v = \frac{m_{EQ} - m_0}{m_0} \cdot 100\,\%. \tag{8}$$

In general, it is assumed that hydrogels based on the structured hydroxyalkyl (meth)acrylates contain 20–40% of water. It has been stated [52] that the amount of sorbed water for such hydrogels depends on the degree of polymeric matrix cross-linking or on the molecular weight of the polymeric grid fragment between nodes.

Hydrogels based on the synthetic copolymers of polyvinylpyrrolidone can be obtained by three methods:

By the method of free-radical thermopolymerization of water-soluble hydroxyalkyl (meth) acrylates in aqueous solution using water-soluble or alcohol-soluble peroxide initiators, at the temperature of 50–70°C. In this case, network structural parameters and water amount in the hydrogel depend on the amount of water in the reaction mass (**Table 4**) [20, 52]

Based on the reactivity of HEMA:PVP composition, a stable hydrogel can be formed in the process of polymerization in water solution when the amount of aqueous is two to three times higher than the mass of composition that forms polymeric matrix. Resulting hydrogel does not release excess of water due to its high sorption ability of PVP-based polymeric matrix and significantly smaller ratio of macrochain crosslinks (**Table 5**). However, under the major excess of water, resulting hydrogel has lower mechanical resistance (**Table 1**).

By the polymerization of PVP-monomer mixture at the room temperature under the effect of iron(II) sulfate in aqueous media, resulting in hydrogel formation (**Table 6**) [52].

Contents of the components for the preparation of membranes (mass parts)				**Water content (%)**	**k·10^4 (m^3/(m^2·h))**	**Permeability coefficient (mole/(m^2·h))**		
HEMA	**PVP**	**H_2O**	**DMSO**			**Sodium chloride**	**Carbamide**	**Sucrose**
100	—	100	—	40	5	80	13	5
80	20	100	—	48	52	181	36	14
80	20	95	5	48	55	193	—	—
80	20	90	10	47	57	212	—	—
80	20	80	20	47	63	240	—	—
80	20	200	—	55	74	234	59	30
80	20	300	—	61	90	263	60	31
70	30	100	—	53	71	232	59	30
50	50	100	—	61	102	274	65	33

Comments: For membranes 1–5, 8, and 9, the luminous transmission factor is 90–96%; membranes 6 and 7 are opaque.

Table 4. Sorption-diffusion properties of hydrogel membranes (δ = 0.2 mm).

Contents of the components (mass parts)				σ (MPa)	ε (%)
HEMA	PVP	H_2O	DMSO		
100	–	100	–	0.53	165
90	10	100	–	0.46	190
80	20	100	–	0.40	235
80	20	200	–	0.38	245
80	20	300	–	0.37	255
70	30	100	–	0.31	270
50	50	100	–	0.22	295
80	20	90	10	0.41	240
80	20	80	20	0.41	245

Comments: DMSO, dimethyl sulfoxide; σ, the ultimate strength of the film; ε, elongation at break of the film.

Table 5. Dependence of mechanical properties of hydrogels on composition content.

By polymerization in bulk of PVP-monomer mixture under the effect of peroxide or iron(II) sulfate initiators followed by swelling of obtained block in water (**Table 7**).

The degree of equilibrium swelling depends on the sorption ability of a copolymeric matrix. Sorption ability of copolymers, which contain in their structure macromolecules of PVP, is much higher than copolymers, based on the separate monomers dissolved in water (**Table 8**) [53].

Water, sorbed in the volume of hydrogel that is based on the monomer system, is in the two forms—filling free intermolecular volume (free water) and solubilized by polar groups in the form of H-complexes and solvated membranes [8, 53]. Water, associated with H-complexes on the polar groups of matrix, transfers into quasicrystal structure, decreasing mobility of water molecules. The higher amount of polar groups is in the polymeric grid; the higher is the water sorption ability of polymer (**Table 9**) [8, 20].

Blend composition (mass parts)		H	P	E	W	k
HEMA	PVP	(MPa)	(%)	(%)	(%)	
90	10	0.101	11	89	48	1.19
80	20	0.099	13	87	52	1.28
70	30	0.082	15	85	63	1.34
60	40	0.079	18	82	69	1.35
50	50	0.050	21	79	76	1.36

Comments: H, hardness number; E, elasticity index; P, plasticity index; W, water content; k, hydrogel swelling factor.

Table 6. Dependence of physical-mechanical properties of copolymers obtained in solution on the blend composition (T = 298 K, $[FeSO_4]$ = 0.01%; blend:H_2O = 1:1).

$[FeSO_4]$ (wt %)	M_C (kg/mol)	ν (mol/kg)	f (%)	p (%)	Copolymer composition (wt %)	
					polyHEMA	PVP
0.01	15.8	0.063	76/91	15/19	84/82	16/18
0.03	14.8	0.067	77/67	15/14	84/86	16/14
0.05	13.6	0.074	79/59	16/13	83/87	17/13
0.07	13.2	0.076	82/37	17/8	83/92	17/8

Comments: numerator, values of block copolymers; denominator, values of copolymers synthesized in water (H_2O:blend = 1:1 w/w); M_c, molecular weight of the chain fragment between polymeric network points; ν, network density; f, grafting efficiency; p, grafting degree.

Table 7. Effect of $FeSO_4$ concentration on the grafting efficiency, grafting degree, and copolymer composition (T = 298 K; HEMA:PVP = 80:20 w/w) [52].

Contents of the components (mass parts)			M_n (kg/mole)	Water content (W, %)	$k \cdot 10^4$ ($m^3/(m^2 \cdot h)$)
HEMA	PVP	H_2O			
100	–	100	12	42	5.1
100 (T3EGDMA)	–	100 (ethanol)	–	1	0.8
80 (GMA)	20	100 (ethanol)	–	20	–
90	10	100	20	45	29.0
80	20	100	24	48	52.3
80	20	200	28	55	74.2
80	20	300	–	61	90.3
70	30	100	38	53	71.4
50	50	100	51	61	102.1
Methylcellulose membrane			–	–	4.0

Comments: k, the permeability coefficient of water; GMA, glycidyl methacrylate; T3EGDMA, triethylene glycol dimethacrylate.

Table 8. Dependence of M_n and W_{H2O} of hydrogels on composition contents.

For PVP sorption of water has specific characteristics.

(a) Hydrate membranes are formed as a result of physical interaction of water with PVP around its elements. In those membranes, due to hydrogen bonds between molecules of water and groups of –N-C = O, redistribution of electronic density occurs that might promote formation of hydroxonium on pyrrolidone cycles. Rothschild [54] offered the scheme of such interaction.

During this interaction of PVP with molecules of water series of changes in pyrrolidone ring occurs.

Content of the components for the preparation of membranes (mass parts)				Water content (W, %)	Coefficient of permeability	
					by water $K \cdot 10^4$ ($m^3/(m^2 \cdot h)$)	by NaCl α ($mole/(m^2 \cdot h)$)
PVA	PVP	HEMA	H_2O			
30	–	30	200	84	52	404
5	25	30	200	75	150	320
10	25	30	200	80	280	320
15	30	15	200	81	280	437
–	–	100	100	40	5	80

Table 9. Sorption-diffusion parameters of hydrogel membranes (thickness 30 μm).

According to the authors [55, 56], about 70% of hydrolyzed rings form hydrogen bonds with water (H-complexes). At the same time, it was found that around such a ring 55 molecules of water are placed in the form of solvated layers—hydrate membranes. The polarization degree of water molecules depends on the distance from ligand-polarized group. Membranes are the least polarized at the external hydrate layers.

(b) Molecules of water that are located on the large distance from carbonate groups of PVP ring do not interact with a ligand: they are kept by the previous membranes with the hydrogen bonds.

(c) Water molecules, kept by hydrophobic fragments of PVP chains, are right next to active complex-forming sites. These molecules can have significant effect on the intermolecular interactions of PVP with additional reagent.

(d) As a result of highly polarized group (ions of hydroxonium), chemical hydration of this group by water molecules can occur [57].

Two percent of pyrrolidone rings can participate in the hydration.

As a result of the high sorption ability of PVP due to numerous physical and chemical interactions with water, it is characterized by significant hygroscopicity. It can sorb and keep large amount of water from the air (**Figure 4**). Moreover, curves of sorption and desorption of water from the air do not match [58].

The desorption curve is at a higher level than sorption curve, indicating a high water-binding power by PVP links, which are characterized by previous interactions (**Figure 5**).

Due to the specificity of the interaction of PVP with water, the coefficient of swelling and water sorption for copolymers on its basis is much higher than those inherent to structured monomer matrices. For PVP copolymers, the swelling coefficient is 1.22…1.35, and the water content is within the range of 47–60% (**Table 10**) [53].

At the same time, it was established that the water sorption and the swelling coefficient practically do not depend on the degree of cross-linking of the polymer matrix (by the amount of dimethacrylate). Water sorption can be the same for both the greater and the smaller cross-link density, if the amount of PVP in the (co)polymer is changed [59].

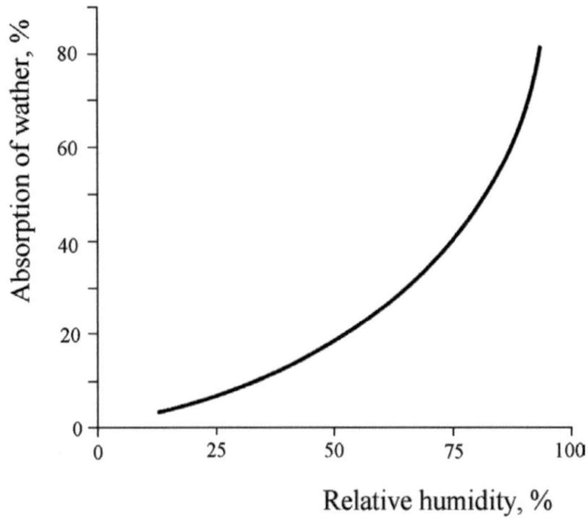

Figure 4. Adsorption of water by PVP from the atmosphere (25°C) for 7 days.

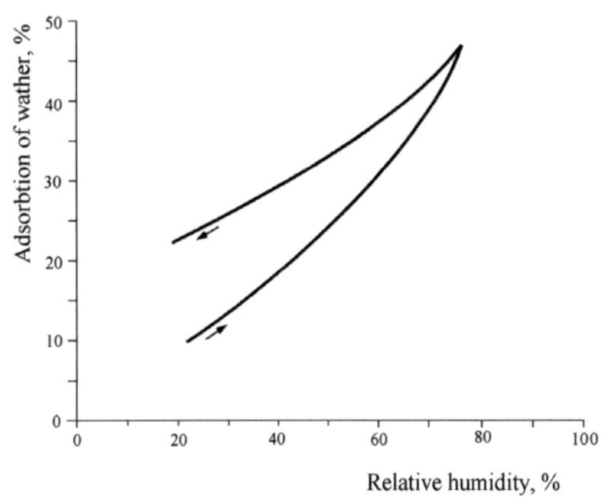

Figure 5. Adsorption and desorption of water by PVP from atmosphere (25°C) [58].

Material of membranes	Heparin sorption (10^{-3} u/m^2)	Heparin desorption for 24 h (%)			$K_{NaCl}\cdot 10^5$ (mole·m^{-2}·h^{-1})
		pH = 2.7	pH = 7	pH = 9.1	
PHEMA	115	4	8	80	212/242*
PVP-gr-PHEMA	550	0	0	2	848/865*
Methyl cellulose	126	8	5	95	–

Comments: δ = 200 μm; K_{NaCl} is a permeability coefficient for NaCl; *for heparinized membranes.

Table 10. Heparin immobilization by membrane surface and their permeability.

4. Application of the practical use of hydrogels based on copolymers of PVP and (meth)acrylates

4.1. Sorption-active granular copolymers of methacrylic acid esters with polyvinylpyrrolidone [60]

Granular copolymerization of 2-hydroxyethyl methacrylate and glycidyl methacrylate with polyvinylpyrrolidone in inert solvents was studied. In suspension (co)polymerization of HEMA with PVP using both PVP and PVA, as stabilizers and also magnesium hydroxide, we obtained spherical particles of satisfactory polydispersity.

The copolymers synthesized are promising as polymer systems for prolonged and controlled drug release. Spherical polymeric particles of size 0.25…2 mm were prepared by suspension copolymerization of the formulations of 2-hydroxyethyl methacrylate and glycidyl methacrylate with polyvinylpyrrolidone. The size and polydispersity of the particles can be controlled by varying the process parameters. The copolymers synthesized exhibit an increased ability to sorb anionic substances, with their subsequent prolonged release in alkaline medium. The composition and particle size of the (co)polymers determine the fields of their application and their performance in prolonged drug release systems.

We researched the effect of the main component ratio of the initial composition on sorption-desorption properties of the granulated polymers based on the results shown in **Figure 6**.

As seen from the obtained results, the lowest observed sorption capacity have homopolymers based on HEMA (**Figure 6**, curve 1). And, efficient sorption has been observed in the first 4 h of the process and continue virtually unchanged. The granulated drug carriers of "Sferogel" provide an effective control of release at a constant rate during the first 8…1 h (**Figure 7**, curve 2).

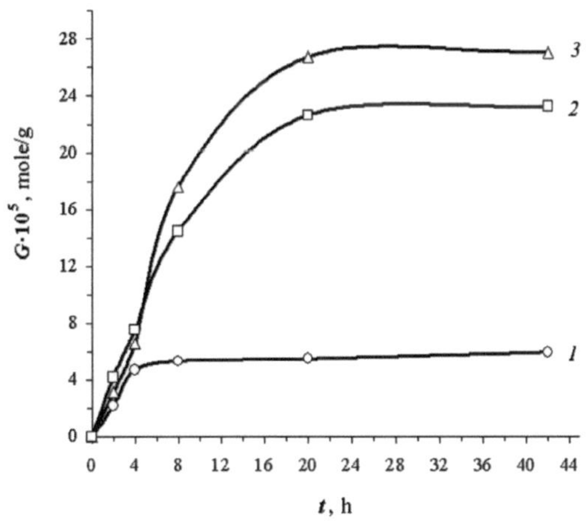

Figure 6. The kinetic curves of diclofenac sodium sorption (G) by granulated hydrogel polymer (SG): HEMA:PVP, wt. p.: (1) 10:0 (SG-1); (2) 8:2 (SG-2); (3) 7:3 (SG-3); $d_{ev.}$ = 640 μm.

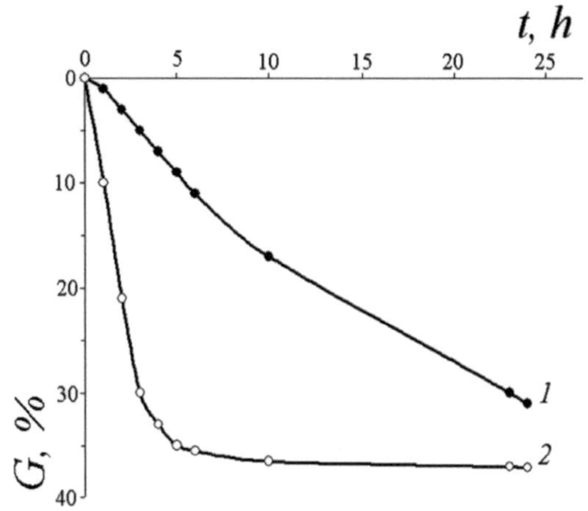

Figure 7. Kinetic curves of diclofenac sodium desorption: (1) for film composite materials filled with SG (HEMA:PVP:SG-2:H_2O = 8:2:1:15 wt. p.) and (2) for SG-2.

If the granules are placed in a hydrogel film, the induction period of 1 h is observed during release when the drugs diffuse from the granules through the film; then, the stable and prolonged release takes place into the environment during the day (**Figure 7**, curve 1).

4.2. High-hydrophilic and thromboresistive dialysis membranes [61, 62]

Development of hemodialysis membranes, cardiovascular implants, and other artificial organs put forward the problem of thromboresistive material creation. One of the effective ways of thromboresistance increase is immobilization of heparin, which is a natural blood anticoagulant, over material surface. The main problem of heparin immobilization by polymeric membranes is its permanent minimal desorption at a contact with blood.

Netted of HEMA/PVP copolymers are perspective compounds for the production of dialysis membranes. The presence of PVP ionic groups in the composition of mentioned copolymers assumes the expansion of biochemical and sorption characteristics and obtaining of membranes with additional functions on their basis.

Hydrogel membranes were obtained by graft polymerization of HEMA over PVP (molecular mass was $10 \ldots 50 \cdot 10^3$) in an aqueous medium, which allowed to combine the synthesis stage and membrane swelling. The saturation of membranes with heparin was realized in glycerol buffer solution (1 M glycerin solution, pH = 2.7), which contained 250,000 units of heparin in 1 l. The amount of sorbed and desorbed heparin was determined by photocolorimetry, based on quantitative determination of heparin and methylene blue complex. Synthesized hydrogel membranes with PVP links have advanced the immobilization ability relative to heparin (**Table 10**).

PVP–heparin complex is so strong, that heparin does not desorb for 24 hours (see Table 10) from the membranes keeping in solutions with different pH (glycin buffer solution with pH=2.7, physiological solution with pH = 7, and solution of sodium tetraboric acid with pH = 9.1). Here, the selective transport characteristics of membranes are changed insignificantly. As for membranes based on polyHEMA and modified cellulose, there is an insignificant precipitation of anticoagulant in acid and neutral media, while in alkaline medium, it grows to 80…95%.

We have established that the presence of –OH and N–C = O hydrophilic groups in the composition of membrane copolymers increases their sorption ability which is characterized with water content (**Table 11**). The increase of PVP content multiplies dialysis permeability (K_{NaCl}) of hydrogel membranes based on HEMA/PVP, but their strength falls down (**Table 11**). Hence, changing hydrogel chemical structure, it is possible to change permeability of membranes on the basis of HEMA/PVP copolymers.

4.3. Hydrogel membranes based on cross-linked copolymers of polyvinylpyrrolidone [63]

At the same time, the hydrogel membranes based on HEMA/PVP copolymers have higher sorption properties compared with HEMA copolymers and higher penetrability for water and several low molecular mass compounds (**Table 12**).

It is interesting that the mass between cross-links does not directly depend upon the solvent polarity. One would expect such a dependence of the given "loosing" effect on the PVP molecules. However, when the solvent amount exceeds its maximum sorption by the polymeric matrix at swelling equilibrium, the already mentioned phase separation occurs (**Table 12**).

Thus, the control of the initial mixture composition via complex formation is an effective method of structure and penetration control for hydrogel membranes based on hydroxyalkyl methacrylates and polyvinylpyrrolidone. Membranes may be recommended for encapsulation and creation of prolonged forms of drug's controlled release and hemodialysis, as well as for fractionating and concentrating of high molecular mass compounds, including biological media.

Contents of the components (mass parts)		**Membrane tensile strength (MPa)**	**Water content (%)**	**K_{NaCl} (mole·m^{-2}·h^{-1})**
HEMA	**PVP**			
100	—	0.53	40	127
91	9	0.46	45	293
82	18	0.40	48	412
77	23	0.31	53	506
69	31	0.22	61	611

Comments: δ = 200 μm; K_{NaCl} is a permeability coefficient for NaCl.

Table 11. Properties of hydrogel membranes based on HEMA/PVP.

Contents of the components (mass parts)				Water content (%)	K·10^4 ($m^3/m^2·h$)	Penetration coefficient* ($mole/m^2h$)		
HEMA	PVP	H_2O	DMSO			NaCl	Carbamide	Saccharose
100	–	100	–	40	5	80	13	5
80	20	100	–	48	52	181	36	14
80	20	95	5	48	55	193	–	–
80	20	90	10	47	57	212	–	–
80	20	80	20	47	63	240	–	–
80	20	200	–	55	74	234	59	30
80	20	300	–	61	90	263	60	31
70	30	100	–	53	71	232	59	30
50	50	100	–	61	102	274	65	33

Comments: *for $\delta = 200$ μm; K, coefficient of water permeability; optical transmission coefficient is 90–96% for experiments 1–5, 8, and 9; opaque membranes have been obtained in experiments 6 and 7.

Table 12. Sorption-diffusion properties of hydrogel membranes.

4.4. Polyvinylpyrrolidone cross-linked copolymers for capsulated particles of drugs [64, 65]

Copolymers synthesized in the form of membranes were effective capsulated agents of solid drugs. In dry state, while storing, they act as protective envelope, but while operation they are able to swell in the physical solution and become permeable. The transferring mechanism of components, including drugs, from encapsulated particles involves several stages (**Figure 8**):

- Swelling of the hydrogel membrane
- Molecular diffusion inside the capsule
- Mass transfer through the hydrogel membrane to the surrounding solution

The used capsule is excreted naturally, without causing any collateral damage to the body.

We also examined the drug release by spherical particles because they model the behavior of prolonged drug while operation.

Thus, we established the relationship between synthesis conditions, structure, and sorption-desorption properties of PVP cross-linked copolymers, what offers their application as carriers for the systems of drug's directional and controlled release.

4.5. Soft contact lenses [66]

It should be noted that the change of the structure and composition of copolymers may considerably influence the size of refraction index n_D. This was consequently used for optimization of

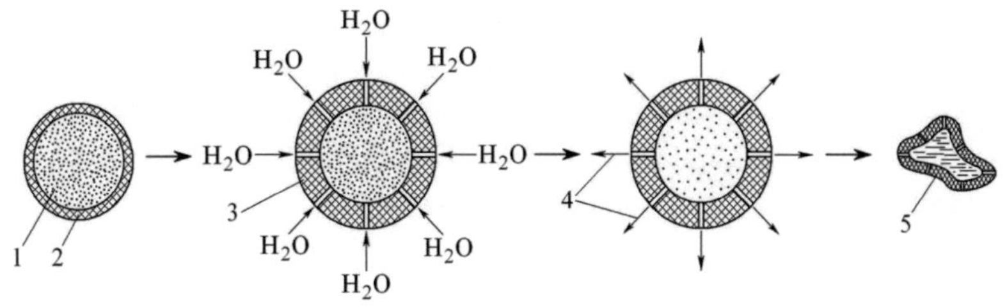

Figure 8. The scheme of component transfer from encapsulated particles: (1) dosage form, (2) hard polymeric shell, (3) swollen hydrogel, (4) release of dosage form, and (5) used capsule.

copolymer composition for contact lenses. It allowed to manufacture correctional soft contact lenses "Akrylan-LPI" with the following operational properties (**Table 13**).

Good permeability for a series of substances, including medicinal solutions, compatibility with alive tissues, and acceptability, has caused the use of the synthesized copolymers for medical ophthalmologic elements of the various geometric shapes. Significant advantage of contact lenses based on PVP copolymer is an essential retention of UV rays and increased oxygen permeability. It provides the lens comfort while long staying on the eye's cornea.

The comparative clinical tests of a condition of an acuteness of vision of an eye without correction and portable spectacle correction were carried out in Lviv Railway Clinical Hospital. From 163 patients without having correction of an acuteness of vision less 0.1 after corrections by contact lenses, an acuteness of vision has increased more than in 80% and has made 0.85–1.0. Researches of a condition of an epithelial integument of a cornea carried out in a various lines after acclimatization at all patients have shown that infringement of integrity of a cornea epithelium does not occur. And, only at six patients after long continuous application of lenses (more than 3 days), mild inflammation of an epithelium was observed.

In this, application of soft hydrophilic contact lenses in treatment of eye diseases is a new promising approach. It substitutes surgical methods in treatment of burns, prevents a symblepharon formation, allows a late keratoplastic, improves results, and decreases treatment duration with high social and economic impact.

Properties in hydrated condition	Parameter meaning
Absorption of water (%)	51
Oxygen permeability ($\times 10^{10}$ $m^2 \cdot s^{-1}$)	1.2
Water permeability ($\times 10^4$ $m^3 \cdot m^{-2} \cdot h^{-1}$)	52
NaCl permeability ($mole \cdot m^{-2} \cdot s^{-1}$)	180
Toughness at a stretching (MPa)	0.4
Relative tensile elongation (%)	250
Permeability of light (%)	96
Refraction index (n_D)	1.4253

Table 13. The characteristics of a polymeric material for soft contact lenses "Akrylan-LPI".

Clinical trial batch of 460 soft contact lens materials of "Akrylan-LPI" in the Laboratory of contact correction of the Filatov Institute of Eye Diseases and Tissue Therapy (Odesa) has been conducted. The comparative study on eye visual acuity, corrected with soft contact lens material "Akrylan-LPI" lenses and contact lens from polyHEMA, has been held on 180 eyes in order to evaluate the optical correction of soft contact lenses.

5. Conclusions

The charge-transfer complex between polyvinylpyrrolidone and 2-hydroxyethyl methacrylate has been determined to affect the polymerizability of PVP/HEMA formulations and the structure of the resulting copolymers. Increasing Kst for the PVP/HEMA complexes has been shown to increase the cross-linking degree of the formed polymer network. It has been revealed that loosely cross-linked PVP/HEMA copolymers and hydrogels based on them can be developed without any radical initiator or in the presence of iron(II) and iron(III) ions. The synthesized hydrogels have increased water content, and their mechanical properties can be easily tuned in a wide range. Moreover, the hydrogels possess high permeability for low-molecular water-soluble substances. Hydrogels also are able for selective sorption of drugs, including a blood anticoagulant heparin. The developed hydrogel materials have been widely tested in industry and recommended for manufacturing various products of medical applications.

Author details

Oleh Suberlyak* and Volodymyr Skorokhoda

*Address all correspondence to: suberlak@polynet.lviv.ua

Lviv Polytechnic National University, Department of Chemical Plastic Processing, Lviv, Ukraine

References

[1] Papkov SP. Studneobraznoe sostoyanie polimerov. Khimiya: Moscow; 1974. p. 256

[2] Sharma KV, Affrossman S, Pethrick RA. Copolymers of 2-hydroxyethyl methacrylate and methyl methacrylate: An electron beam resist study. Polymer. 1984;**25**(8):1090-1092. DOI: 10.1016/0032-3861(84)90344-6

[3] Seiderman M. Hydrophilic gel polymers of vinylpyrrolidone: Patent USA 3721657; 1973

[4] Lavrov NA. Osobennosti polucheniya rastvorimyih polimerov na osnove 2-oksietilmetakrilata. In: Khimicheskaya Tehnologia, Svoystva i Primenenie Plastmass. Leningrad: Khimiya. 1983. pp. 14-18

[5] Arbuzova IA. Andreeva GA. Sintez i svoystva polimerov monoakrilata etilenglikolya: Plasticheskie massyi. 1982;**6**:46-47

[6] Shultz Herman. Hydrophilic copolymer: Patent USA 4067839; 1978

[7] Kaetsu Isaa, Kumakura Minoru, Ito Okhito. Soft contact lenses and process for preparation thereof: Patent USA 3983083; 1976

[8] Refojo MF, Yasuda H. Hydrogels from 2-hydroxyethyl methacrylate and propylene glycol monoacrylate. Journal of Applied Polymer Science. 1965;**9**(7):2425-2435. DOI: 10.1002/app.1965.070090707

[9] Migliaresi C, Nicodemo L, Nicolais L, Passerini P, Stol M, Hrouz J, Cefelin P. Water sorption and mechanical properties of 2-hydroxyethyl-methacrylate and methylmethacrylate copolymers. Journal of Biomedical Materials Research. 1984;**18**(2):137-146. DOI: 10.1002/jbm.820180204

[10] Gustatson R. Copolymerized hard plastic hydrogel compositions: Patent USA 3728315; 1973

[11] Seiderman M. Improvements in relation to the preparation of hydrophilic gel polymers: Great Britain Patent 1339727; 1974

[12] Gustatson R. Copolymerized hard plastic hydrogel compositions: Patent USA 3892721; 1975

[13] Howes J., Gordon B., Selway A. Cross linked polymer: Great Britain Patent 1494641; 1977

[14] Rostoker M, Levine L. HEMA-copolymers: Patent USA 4038264; 1977

[15] Neogi Amar N. Hydrophilic polymer composition for prosthetic devices: Patent USA 3876581; 1975

[16] Ubbach J. Hydrophilic contact lens material. Great Britain Patent 38516176; 1978

[17] Loshaek S. Contact lenses of high water contact. Patent USA 4158085; 1979

[18] Carle Trevonde. Hydrophilic gel. Patent USA 3937680; 1975

[19] Crosslinked polymer. Great Britain Patent 1514810; 1978

[20] Seiderman N. Hydrophilic polymer: Patent USA 3792028; 1974

[21] Taruni Niro, Tuchia Makoto. Process of producing soft contact lenses: Patent USA 4143017; 1977

[22] Seiderman N. Contact lenses: Patent USA 3767831; 1974

[23] Masuhara Eyiti. Preparation of soft contact lenses: Japan Patent 54-3733; 1979

[24] Wingler F, Leuner B, Schwabe P. Kontactlinsen aus Methacrylsaueremethylester. Copolymerisaten: Germany Patent EP0027221 B1; 1983

[25] Starodubtsev SG, Georgieva VR, Pavlova NRO. korrelyatsii vraschatelnoy podvizhnosti spinovogo zonda i kislorodnoy pronitsaemosti gidrogeley: Sinteticheskie polimeryi meditsinskogo naznacheniya. In: Materialy IV Vsesoyuznogo Nauchnogo Simpoziuma. Dzerzhinsk: USSR; 1979. pp. 13-15

[26] Krasinskyi VV, Antoniuk VV, Yakhovich T, Vasyshak RI. Ekspluatatsiini vlastyvosti plivok na osnovi polivinilovoho spyrtu ta modyfikovanoho montmorylonitu. Visnyk NU "Lvivska politekhnika": Khimiia. Tekhnolohiia Rechovyn ta Yikh Zastosuvannia. 2016;**841**:377-383

[27] Suberlyak OV, Zaikina OS. Makromolekuliarnyi initsiator polimeryzatsii (met)akrylativ kompleksnoho typu. Dopovidi Akademiyi Nauk URSR B. 1990;**11**:53-56

[28] O'Driscol K., Isen A. Fabrication of soft plastic contact lens: Patent USA 3841985; 1975

[29] Ericson E, Neogi A. Hydrophilic polymers and devices made there form: Great Britain Patent 1412439; 1975

[30] Neeta T, Srivastava AK. Poly(2-vinyl pyridine) as a template for the radical polymerization of methyl acrylate. Indian Journal of Chemistry. A. 1990;**29**(4):324-327

[31] Evelle D. Hydrophilic contact lens material: Patent USA 3647736; 1972

[32] Le Boeuf A, Grovesteen W. Pyrrolidone-methacrylate graft copolymers from 3-stage process: Patent USA 3978164; 1976

[33] Suberlyak OV, Levitskij VE, Skorokhoda VY, Godij AB. Physical-chemical fenomena on phase boundary vinyl monomer-water solution of polyvinylpyrrolidone. Ukrainskij Khimicheskij Zhurnal. 1998;**64**(6):122-125

[34] Sidelkovskaya F.P. Himiya N-vinilpirrolidona i ego polimerov. Moscow: Nauka; 1970. 160 p

[35] Gallop P. Korb D. Polymeric compositions and hydrogels formed therefrom: Patent USA 4379864; 1983

[36] Hrytsenko OM, Suberliak OV, Moravskyi VS, Haiduk AV. Doslidzhennia kinetychnykh zakonomirnostei khimichnoho osadzhennia nikeliu. Skhidno-Yevropeiskyi Zhurnal Peredovykh Tekhnolohij. 2016;**1/6**(79):26-31

[37] Levytskyi VY, Hancho AV, Suberlyak OV. Fizyko-khimichni zakonomirnosti formuvannia polivinilpirolidon-sylikatnykh nanokompozytsiinykh materialiv. Voprosyi Himii i Himicheskoy Tehnologii. 2010;**6**:55-59

[38] Frömming KH, Ditter W, Horn D. Sorption properties of cross-linked insoluble polyvinylpyrrolidone. Journal of Pharmaceutical Sciences 1981;**70**(7):738-743. DOI: 10.1002/jps.2600700707

[39] Plazier-Vercammen JA, De Nève RE. Interaction of povidone with aromatic compounds. I. Evaluation of complex formation by factorial analysis. Journal of Pharmaceutical Sciences. 1980;**69**(12):1403-1408. DOI: 10.1002/jps.2600691213

[40] Gustavson KH. Note on the fixation of vegetable tannins by polyvinylpyrrolidone. Svensk Kemisk Tidskrift. Stockholm. 1954;**66**:359-362

[41] Schenck HU, Simak P, Haedicke E. Structure of polyvinylpyrrolidone-iodine. Journal of Pharmaceutical Sciences 1979;**68**(12):1505-1509. DOI: 10.1002/jps.2600681211

[42] Horn D, Ditter W. Chromatographic study of interactions between polyvinylpyrrolidone and drugs. Journal of Pharmaceutical Sciences 1982;**71**(9):1021-1026. DOI: 10.1002/jps.2600710917

[43] Suberlyak OV, Skorokhoda VI,Thir IG. Complex-formation effect on polymerization of 2-oxyethylene methacrylate in the presence of polyvinylpyrrolidone. Vysokomolekulyarnye soedineniya. B. 1989;**31**(5):336-340

[44] Suberlyak O, Melnyk J, Skorokhoda V. Formation and properties of hydrogel membranes based on cross-linked copolymers of methacrylates and water-soluble polymers. Engineering of Biomaterials. 2009;**12**(86):5-8

[45] Suberlyak O, Skorokhoda V, Grytsenko O. *complex* PVP-Me^{n+}—Active catalyst of vinyl monomers polymerization. In: Materiały Polimerowe i ich Przetwórstwo. Częstohowa: Wydawnictwo Politechniki Częstohowskiej; 2004. p.140-145

[46] Suberlyak OV, Gudzera SS, Skorohoda VI. Osobennosti polimerizatsii 2-oksietilen(met) akrilatov v polyarnykh rastvoritelyakh v prisutstvii polivinilpirrolidona. Doklady Akademii Nauk USSR. B. 1986;**7**:49-51

[47] Suberlyak OV, Mel'nyk YY, Skorokhoda VI. Regularities of preparation and properties of hydrogel membranes. Materials Science. 2015;**50**(6):889-896. DOI: 10.1007/s11003-015-9798-8

[48] Suberlyak OV, Kopel'tsiv YA. Effect of the charge transfer complex in the initiator-free polymerization in the solution of composition of 2-hydroxyethylene methacrylate with polyvinylpyrrolidone. Ukrainskij Khimicheskij Zhurnal. 1993;**2**:213-216

[49] Suberlyak OV, Zaikina OS. Makromolekuliarnyi initsiator polimeryzatsii metakrylativ kompleksnoho typu. Dopovidi Akademiyi Nauk URSR. B. 1990;**11**:53-56

[50] Suberlyak OV, Zaikina OS, Thir IG, Soshko AI. Modifitsirovanie vodorastvorimykh polimerov i gidrogelnye membrany na osnove produktov sinteza. Plasticheskiye Massy. 1985;**11**:27-29

[51] Tanaka Keyti. Polymer with a High Water Content: Japanese Request 52-84273; 1977

[52] Suberlyak O, Grytsenko O, Kochubei V. The role of $FeSO_4$ in the obtaining of polyvinylpyrrolidone copolymers. Chemistry & Chemical Technology. 2015;**9**(4):429-434

[53] Suberlyak OV, Skorohoda VI. Sopolimery (met)akrilovykh efirov glikoley s polivinilpirrolodonom dlya polucheniya vysokopronitsaemykh membran. Zhurnal Prikladnoy Khimii. 1989;**6**:1330-1333

[54] Rothschild WG. Binding of hydrogen donors by peptide group of lactams. Identity of the interactions sites. Journal of the American Chemical Society. 1972;**94**(25):8676-8683. DOI: 10.1021/ja00780a005

[55] Molyneux P, Frank HP. The interaction of Polyvinylpyrrolidone with aromatic compounds in aqueous solution. Part II.1 the effect of the interaction on the molecular size of the polymer. Journal of the American Chemical Society. 1961;**83**(15):3175-3180. DOI: 10.1021/ja01476a002

[56] Süvegh K, Zelkó R. Physical aging of poly(vinylpyrrolidone) under different humidity conditions. Macromelecules. 2002;**35**(3):795-800. DOI: 10.1021/ma011148l

[57] Maruthamuthu M, Reddy JV. Binding of fluoride onto poly(N-vinyl-2-pirrolidone). Journal of Polymer Science. Part C: Polymer Letters. 1984;**22**(10):569-573. DOI: 10.1002/pol.1984.130221012

[58] Sadek HM, Olsen JL. Determination of water-adsorption isotherms of hydrophilic polymers. Journal Pharmacy Technolology. 1981;**5**(2):40-48

[59] Thir IG, Sheketa ML, Zaikina OS, Shulman MS. Zavisimost koeffitsienta nabuhaniya polimerov dlya myagkih kontaktnyih linz ot kompozitsionnogo sostava i rezhima polimerizatsii. Vestnik Lvovskogo Politehnicheskogo Instituta. 1982;**163**:43-45

[60] Suberlyak OV, Semenyuk NB, Dudok GD, Skorokhoda VI. Regular trends in synthesis of sorption-active granular copolymers of methacrylic acid esters with polyvinylpyrrolidone. Russian Journal of Applied Chemistry. 2012;**85**(5):830-838. DOI: 10.1134/s1070427212050254

[61] Skorokhoda V, Melnyk Y, Semenyuk N, Ortynska N, Suberlyak O. Film hydrogels on the basis of polyvinylpyrrolidone copolymers with regulated sorption-desorption characteristics. Chemistry & Chemical Technology. 2017;**11**(2):171-174. DOI: 10.23939/chcht11.02.171

[62] Suberlyak O, Melnyk J, Baran N. High-hydrophilic membranes for dialysis and hemodialysis. Engineering of Biomaterials. 2007;**63-64**(10):18-19

[63] Skorokhoda V, Yu M, Semenyuk N, Suberlyak O. Structure controlled formation and properties of highly hydrophilic membranes based on polyvinylpyrrolidone copolymers. Chemistry & Chemical Technology. 2012;**6**(3):301-305

[64] Suberlyak O, Skorokhoda V, Semenyuk N. The structure and immobilization activity of polyvinylpyrrolidone cross-linked copolymers. Engineering of Biomaterials. 2007;**63-64**(10):14-15

[65] Skorokhoda V, Melnyk Yu, Shalata V, Skorokhoda T, Suberliak S. An investigation of obtaining patterns, structure and diffusion properties of biomedical purpose hydrogel membranes. Eastern-European Journal of Enterprise Technologies. 2017;**1**(6/85):50-55. DOI: 10.15587/1729-4061.2017.92368

[66] Suberlyak O, Skorokhoda V, Kozlova N, Melnyk Yu, Semenyuk N, Chopyk N. The polyvinylpyrrolidone graft copolymers and soft contact lenses on their based. Science Rise. 2014;**5**(5/3):52-57. DOI: 10.15587/2313-8416.2014.33235

Chapter 3

Superabsorbent

Mohamed Mohamady Ghobashy

Additional information is available at the end of the chapter

http://dx.doi.org/10.5772/intechopen.74698

Abstract

Superabsorbent hydrogel (SAH) is a cross-linked polyelectrolyte polymer that has the capability to absorb a lot of water by keeping it in a three-dimensional (3D) structure. The network's structure of SAH has the high elasticity that gives the ability of pores to expand in an aqueous media into up to 150–1500 times their own size in a dry state. The size of pores is the major factor that controls the swelling degree of the hydrogel. In contrast, the swelling degree is related to cross-linked density and the number of polarizable functional groups that immobilize on the polymer backbone. The hydrogels could be made by radical-initiated polymerization of hydrophilic monomers, and/or linear polymers dissolve in an aqueous solution. Free radical polymerization of the hydrogel can be done physically or chemically. Advantages and disadvantages of each method will be elaborated in this chapter. The advances in radiation cross-linking methods for the hydrogel preparation are particularly addressed besides other different techniques, e.g., (freezing/thawing and chemical initiation). This chapter will review the preparation methods of superabsorbent hydrogels from synthetic and natural hydrophilic polymers with other new phases such as wax, gum, and rubber. Methods to characterize these hydrogels and their proposed applications (internal curing agent for cement, agricultural proposal, biomedical proposal, and environmental proposal) are also reviewed.

Keywords: SAH, swelling, cross-linking, polyelectrolyte, hydrogel

1. Introduction

A cross-linked polyelectrolyte polymer that has the capability to absorb water up to 150–1500 times of their own size in a dry state and keep water in a three-dimensional (3D) structure is called superabsorbent hydrogel (SAH) (**Figure 1**). Several descriptions are used to express superabsorbent hydrogel (SAH): (I) SAH has synthetic three-dimensional swollen networked structures. The product from covalently cross-linked to certain synthetic

Figure 1. Hydrogel is a hydrophilic monomer that would be cross-linked.

monomer or nature polymer [1–3]. (II) SAH is a fine white powder-like sand or tiny granule-like sugar and has a high ability of water absorption. (III) SAH is the diffusion of the solvent (typically water) into the hydrogel network when the hydrogel is placed in contact with water; the water molecules begin to diffuse inside the solid hydrogel network. (IV) By another way, at equilibrium swelling hydrogel contains small fractions of solid and large fractions of water so it can be described as a hydrogel that begins to diffuse into the water. The diffuser of the water outlet of the hydrogel is a deswelling process. Over time, when the maximum swelling of hydrogel soaked in water is achieved, the bonds of the network will relax by evolving apportion of the water molecules (deswelling) and absorb of water molecules again (swelling) with the time interval. This deswelling/swelling process affecting by the pressure of water molecules on the bonds of hydrogel networks that given the swelling curve line saw shape (zigzag), this phenomenon so-called hydrogel breath. **Figure 2** is a representation of the 3D structure of a swollen and dried hydrogel.

From **Figure 2** it can be said that the hydrogel chains are in close and function groups tightly interacting with each other due to H bonds. As water diffuses inside the hydrogel network, the function groups begin to hydrate, and the interactions such as H bonds will terminate. With further water molecules absorbed, the chains will gain pressure with a gain swell. At appropriate conditions, the hydrogel will reach a state where the pores are fully filled with water and chains reach the maximum expanded. Responding to many external stimulus conditions, expansion and shrinkage of hydrogel are controlled [4].

2. Kinetics of water absorbed (swelling) in hydrogel

The swelling process of a hydrogel is a transition from solid state to fluid state without dissolution or dissociation. The two interfaces become one interface called "gel." Eq. (1) was used to determinate the nature of diffusion of water molecules into the hydrogels:

http://dx.doi.org/10.5772/intechopen.74698

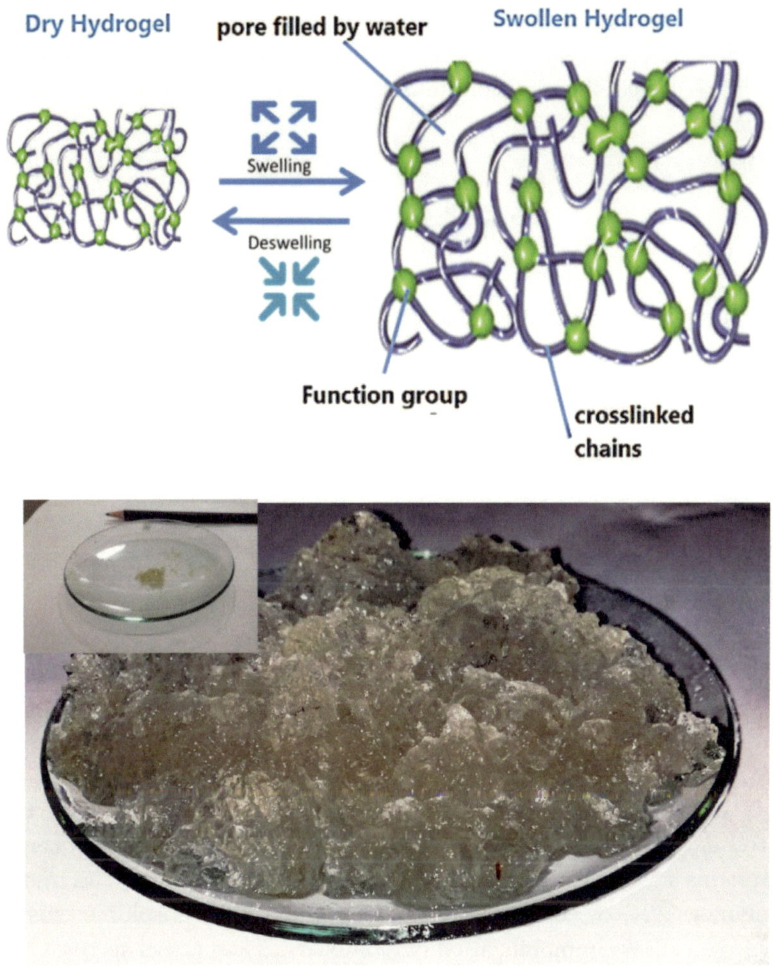

Figure 2. Illustration in the left hydrogel in a dry state or deswelling state (shrinking).

$$\frac{M_{t'}}{M_{\infty'}} = K\,t^{n} \quad (1)$$

where M_t and M_∞ are the amount of water diffusion into the hydrogel at the time (t); at the infinite time, respectively, K is a constant related to the structure of the network; and n is a characteristic exponent of the transport mode of the water solvent [5]. Depending on the relative rates of water diffusion and hydrogel network relaxation, three cases of diffusion mechanisms are distinguished. The Fickian diffusion may be described by Case I which appears when the T_g of hydrogel is below the water medium temperature. In this case, the hydrogel chains have a high mobility and relaxation, and the water penetrates more easily into the relaxed network. Therefore, the water diffusion rate, R_{diff}, is much less than the hydrogel chain relaxation rate R_{relax} ($n = 0.50$) ($R_{diff} \ll R_{relax}$). Case II is the non-Fickian diffusion, in which diffusion is very rapid compared to the relaxation processes ($0.50 < n < 1$), which appears when the T_g of hydrogel is well above the experimental temperature. In this situation, the hydrogel chains are not adequately mobile to permit urgent penetration of water into the hydrogel ($R_{diff} \gg R_{relax}$). Case III is the anomalous diffusion. It is observed when the diffusion and relaxation rates are comparable ($R_{diff} \approx R_{relax}$) [6]. To detect the diffusion mechanisms, the swelling curves are fitted to Eq. (1) which becomes:

$$\log\left(\frac{M_t}{M_\infty}\right) = logK + n\log t \quad (2)$$

The diffusional exponent n is calculated from the slopes and K (kinetic rate of swelling) from the intercept.

3. Swelling properties of superabsorbent hydrogels

3.1. Natural polymer

The natural polymers copolymerized with synthetic polymers provided these natural polymers to have a suitable functional group and gain mechanical strength more than individual [7]. A variety of natural polymers such as chitosan, heparin, pectin, chitin, hyaluronic acid, agarose, dextran, and alginate are excellent to covalently cross-link and polymerize to form hydrogel. They have been explored as biocompatible, biodegradable hydrogels for biomedical applications [8–13]. Polysaccharide hydrogel is a biopolymer with high permeability for nutrients, oxygen, and other water-soluble metabolites, making it attractive scaffolds for use in cell encapsulation [14–16]. In addition, polysaccharides have been copolymerized with proteins such as laminin, gelatin, collagen, and fibrin to form an interpenetrating network polymer (IPN) or composite hydrogels [17–21]. This insoluble cross-linked biopolymer hydrogel allows immobilization of biomolecules and active agents. Owing to their high water retender, hydrogels resemble natural soft tissue more than any other type of polymeric biomaterials [22]. The water retention in SAH-based natural polymer promotes cell migration, growth, proliferation, differentiation, and adhesion, leading to tissue regeneration scaffolds [23].

3.2. Gum and wax

Hydrocolloid materials such as wax and gum and a few others such as surfactant or acidic oils have been copolymerized with hydrogel and novel systems of hydrocolloid hydrogel matrices with attractive properties. Emulsion polymerization is an efficient method for the production of new wax-hydrogel such as cetyl alcohol: stearic acid-based acrylamide hydrogel by using triethylamine (TEA) as an emulsifier [24]. A cross-linking reaction is performed at a dose of 20 kGy. This wax-hydrogel matrix bearing of acid and amide groups shows high swelling value behavior at different pH values.

Figure 3 shows the SEM images of wax-hydrogel matrices of CtOH-StA/PAAm. It was observed that the perfect miscibility was between wax and hydrogel networks as one phase with the absence of a separation zone. **Figure 3a** shows the equilibrium swelling at pH 6

http://dx.doi.org/10.5772/intechopen.74698

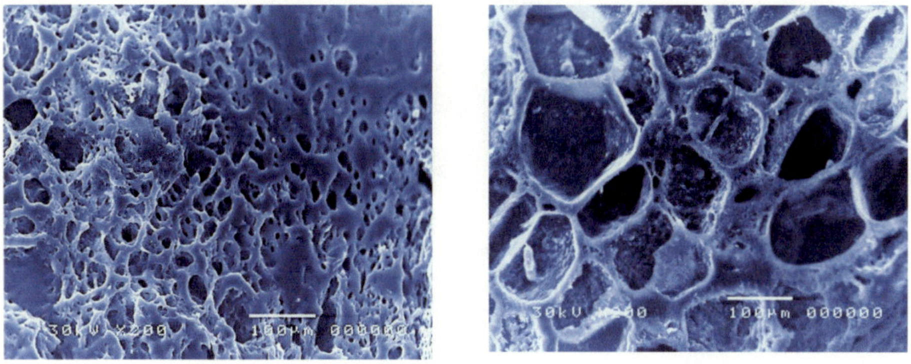

(a) (b)

Figure 3. Scanning electron microscopy (SEM) images show the porous structure of CtOH-StA/PAAm (a) swelled at pH 6 and (b) swelled at pH 10.

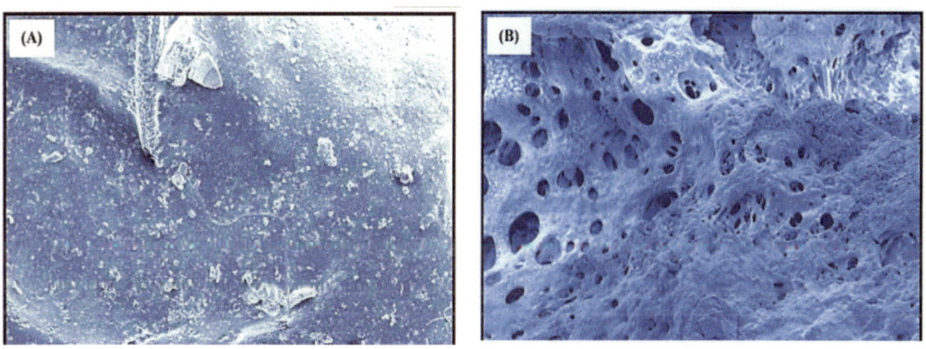

Figure 4. Scanning electron microscopy (SEM) images of (PAAm-g-XG) (A) swelled at pH 1.2 and (B) swelled at pH 7.4.

and the interconnected porous structure. At pH 10 (**Figure 6b**), an alveolate morphology and highly uniform pores are observed.

Another example of pH-sensitive gum (xanthan) based on acrylamide hydrogel is given (PAAm-g-XG) by radical polymerization [25]. **Figure 4** shows SEM images of the pH-sensitive swelling behavior reflecting the surface morphology of (PAAm-g-XG) exposed to acidic and alkaline pH. **Figure 4A** at pH 1.2 surface of (PAAm-g-XG) shows no pores as there was minimum swelling. In **Figure 4B**, at pH 7.4, the surface morphology of PAAm-g-XG reveals highly porous structure compared with the surface of PAAm-g-XG incubated in pH 1.2. As shown above, the swelling in alkaline medium is higher than in the acidic one due to the $CONH_2$ in acrylamide groups and hydrolysis to COO^- affected by NaOH. The electrostatic repulsion of the ionized groups COO^- leads to increase the pore size.

3.3. Rubber

Rubber could be used for improving the elasticity of many materials due to their flexibility and softness that has glassy transition temperatures bringing down the ambient temperature

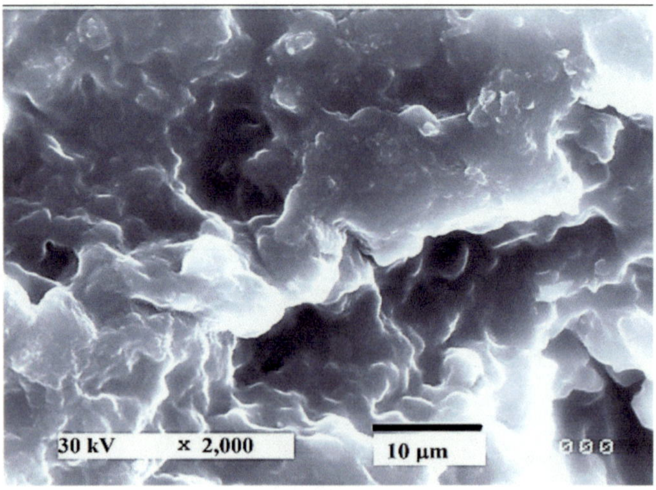

Figure 5. SEM image of SBR/PVP/MAA. The porous structure affected by elasticity of SBR chains.

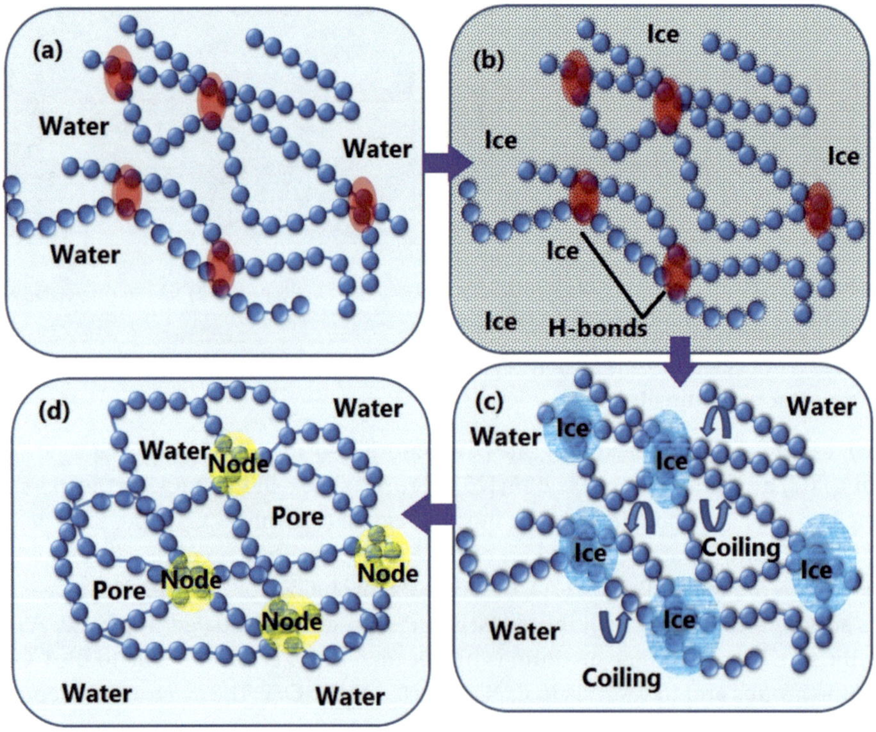

Figure 6. The mechanisms for PVA hydrogel formation from PVA solution by a freezing–thawing method: (a) PVA-water system. (b) PVA-ice system and frozen of chains. (c) PVA-ice-water system obtained by the gradual thawing of the PVA solution causing coiling of PVA chains together to form node (helix hydrogel). (d) Cross-linked network of PVA hydrogel was obtained.

[26, 27]. Water-swollen composite rubber-hydrogel materials are highly permeable to various applications. For example, rubber-hydrogel of SBR/PVP/MAA composite was prepared by mixing styrene-butadiene rubber (SBR) and copolymer hydrogel (polyvinylpyrrolidone-co-methacrylic acid) cross-linked by gamma irradiation. A high miscibility was observed between the MAA/PVP hydrogel and the matrix of SBR with swelling degree of 25 (g/g) after 4 h. **Figure 5** shows a SEM image of SBR/PVP/MAA revealed to a uniform of surface morphology indicating a high compatibility between the hydrogel and rubber matrices [28].

4. Preparation methods of SAH (freezing/thawing, ionizing radiation, and chemical initiation)

Superabsorbent hydrogels are prepared from either natural or synthetic polymers. The synthetic polymers are mechanically robust and stiffer compared to natural polymers. Their mechanically strength results in a slow degradation rate providing the durability as well, but mechanical strength of natural polymers is highly poor. These two inverse properties should be adjusted through ideal design [29]. The following section describes the physical and chemical methods reflecting the synthesis of hydrogels.

4.1. Freezing: thawing process

Freezing–thawing process is a physical method for creating a strong and highly elastic hydrogel [30]. The general advantage of physically forming hydrogel is the need for the addition of cross-linking and initiator entities. PVA is a common polymer that suitable cross-links by freezing–thawing techniques. Cross-linked PVA hydrogel is obtained when an aqueous solution of PVA highly crosslinked by the "repeated freezing-and-thawing cycles." The steps in this method were as follows. First, an aqueous solution of PVA was frozen at a low temperature 0°C, and then the frozen PVA solution stands to thaw at room temperature or at a temperature of 60°C. These cycles were repeated, with increasing the cycle of freezing the resultant PVA hydrogel with much cross-linked density, and the water resistance has also been increased [31]. **Figure 6** reveals why the "freezing–thawing" process is well involved in the preparation of hydrogel for the PVA aqueous solution [32]. **Figure 6a** shows the H bond formation of PVA chains dissolved in water. Once the PVA solution was frozen (**Figure 6b**), ice crystals were formed within all PVA molecules, and the chains were freezing (low motion). During the "thawing" process, the ice will melt gradually from free chains before bonding chains (H bonds). A free space with water allowing free chains to cling together (tangle) and form node is seen on the left of **Figure 6c**. The "repeated freezing-and-thawing" method cross-linked PVA hydrogel with superporous structure that is formed (**Figure 6d**).

4.2. Chemical cross-linkage

Moreover, physically cross-linked hydrogels have limitations in which a few kinds of polymeric materials could be cross-linked by this method. Chemically, cross-linkage could be carried out in the wide kinds of polymeric matrials in thr presence of initiators and

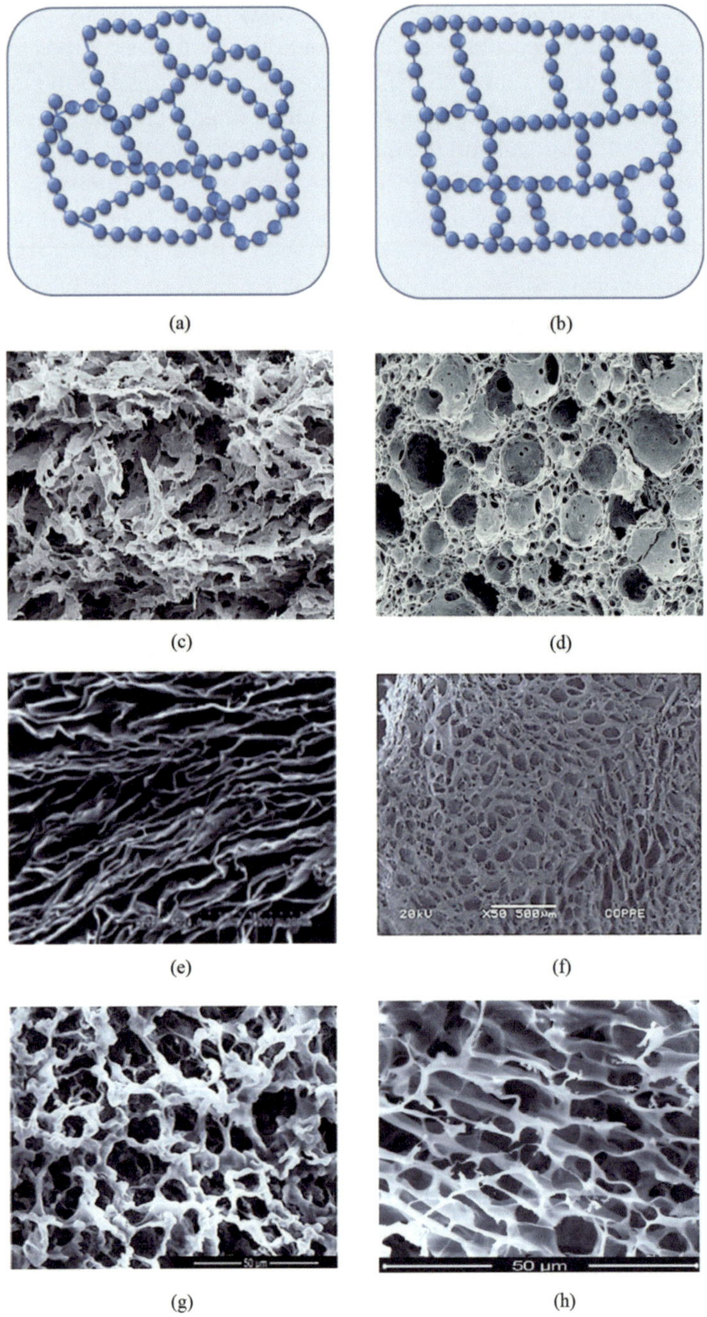

Figure 7. SEM images of morphologies of synthetic hydrogels. The changes in cross-linking method lead to the spontaneous formation of various porous structures. Physical hydrogels are not homogeneous, since clusters of molecular entanglements, or hydrophobically or ionically associated domains, can create homogeneities. Free chain ends or chain loops also represent transient network defects in physical gels. Physical hydrogel (a) chemical hydrogel (b) PVA hydrogel crosslinked (c) physically and (d) chemically (e) and (f) shows the chitosan hydrogel crosslinked physically as tubular porous structure and chitosan hydrogel crosslinked chemically using (g) succinic acid and (h) glutaric acid.

http://dx.doi.org/10.5772/intechopen.74698

cross-linking agents. In another known difference, the physically cross-linked hydrogels are not homogeneous networks, the chain entanglements forming irregular porous structure, while chemically cross-linked hydrogels are covalently cross-linked networks forming regular porous structure [33]. **Figure 7** shows the morphologies of synthetic PVA hydrogels by two different processes (physically and chemically). The change in the parameters during cross-linking leads to the spontaneous formation of various porous structures. **Figure 7c** shows SEM image of PVA hydrogel prepared which underwent eight freezing–thawing cycles [34] given irregular porous structure. While **Figure 7d** shows a SEM image and PVA cross-linked chemically by adding H_3BO_3 [35], as seen in **Figure 7d**, the porous structure is more regular compared with the one in **Figure 7c**. **Figure 7e** and **f** shows an image of tubular chitosan prepared using the freezing–thawing method [36, 37]. **Figure 7g** and **h** shows SEM images of the chitosan cross -inked chemically by adding succinic acid [38] and by adding glutaric acid [39], respectively. Open-pore structure with a high degree of interconnectivity can be observed. It is indicated that the mechanism of cross-linking accompanied by reaction-induced phase separation leads to diverse morphologies of the resulting porous structure of cross-linked hydrogel.

4.3. Ionizing radiation

Cross-linking by irradiation occurs using a high-ionizing energy, such as gamma rays (Co-60), x-ray, or electron beam (*e*-beam). Gamma irradiation is more economically rather than the rest of the other irradiation techniques [40, 41]. Gamma irradiation is a promising technique to fabricate a wide scale of different materials especially polymeric materials [42–44]. Particularly, at the first step, a polymer radical is formed with regard to water radiolysis by ionizing radiation follow-up polymerization reactions. Water radiolysis generates six reactive species ($^{\bullet}H$, $^{\bullet}OH$, e_{hy}, H_2, H_2O_2, and $O^{\bullet}$). All $^{\bullet}H$ and $^{\bullet}OH$ radicals beside e_{hy} produced upon radiolysis of water are transferred by the polymeric solute to form carbon chain macroradicals. The average number of macroradical centers formed in a pulse would then be determined simply from the radiation chemical yield of hydroxyl radicals and the dose per pulse. For example, radiation cross-linkage of acrylic acid (AAc) monomer and polymer radicals preferably undergoes intra-cross-linking and inter-cross-linking reactions producing a porous structure. Macroporous structures are produced when inter-cross-linking predominates and intra-cross-linking would cause nano-porous structures [45]. Radicals are generated from the radiolysis of AAc aqueous solutions; the predominance of the inter-cross-linking reactions is achieved due to the large number of carbon–carbon double bonds. The beginning of intra-cross-linking under these conditions is confirmed by the macroporous formation. Also, higher yield of C-centered free radicals along the PAAc chain enhances the intra-cross-linking reactions. Dimmers are formed by combination of two macroradical molecules. In the same manner, a 3D cross-linked hydrogel will be obtained of PAAc polymer. **Figure 8** shows covalently cross-linked hydrogels. First, macroradicals are combined together, and then cellular structure consisted of small compartments or pores as rooms are filled with water. The network expansion probably takes place by means of an own pores are filled with water. At a certain water pressure bonds of the networks, walls contract and relax.

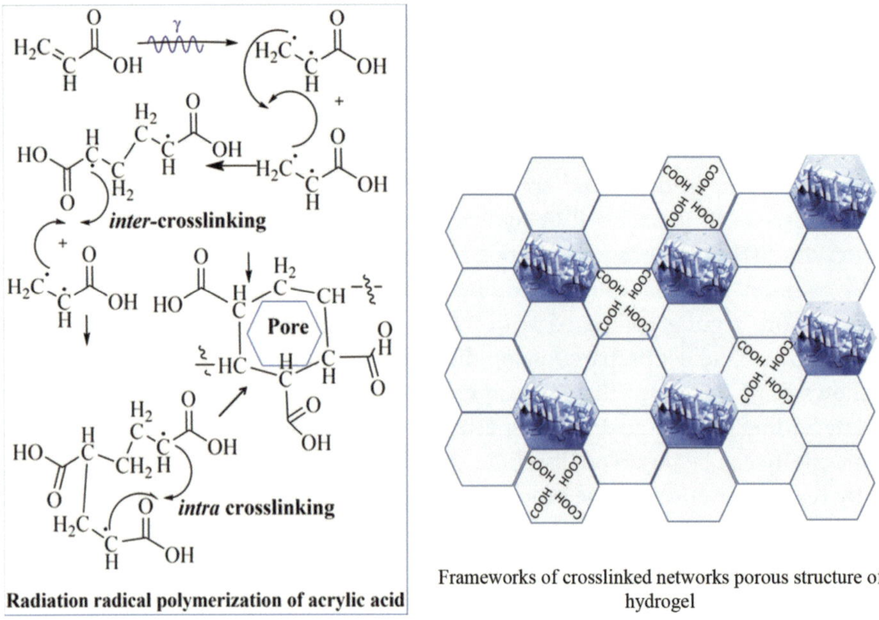

Figure 8. (Left) The proposed radical polymerization mechanism induced by gamma irradiation on the preparation of polyacrylic acid. (Right) The framework of SAH network structure has a preferential spatial orientation.

5. Design of the polymerization reaction according to appropriate monomer

A construction of 3D frameworks of superabsorbent hydrogel is random and amorphous of nature. But this does not prevent inspiration for the molecular architecture of hydrogels. The fabrication of various architectures for SAH is demonstrated in **Figure 9a** and **b** According to the IUPAC definitions [46], semi-IPN networks are composited of one linear polymer entrapped within the network of another polymer, while IPN networks comprise more than one polymer network structures interlaced on a molecular scale. **Figure 9c** is an image of a multilayer hydrogel (MLH) [47] in which single networks are stacked one onto the other, interpenetrating network films with the mixture of two networks in the same layer. **Figure 9f** is an image of hydrogel fibers [48–50]. The synthesis of hydrogel depends on the concentration of monomer, initiator, and cross-linked agents.

The polymerization reaction may occur by two growth polymerizations such as step and chain growth polymerizations individually or of both [51]. In case of chain growth, the polymerization reaction is very fast, whereas in step growth, the reaction is gradual and monomer concentration decreases gradually. In case of chain growth, the polymerized hydrogel conversion is high, while pore size and swelling are low. Whereas in step growth the hydrogel conversion is poor, pore size and swelling are high [52]. **Figure 10** shows the framework

http://dx.doi.org/10.5772/intechopen.74698

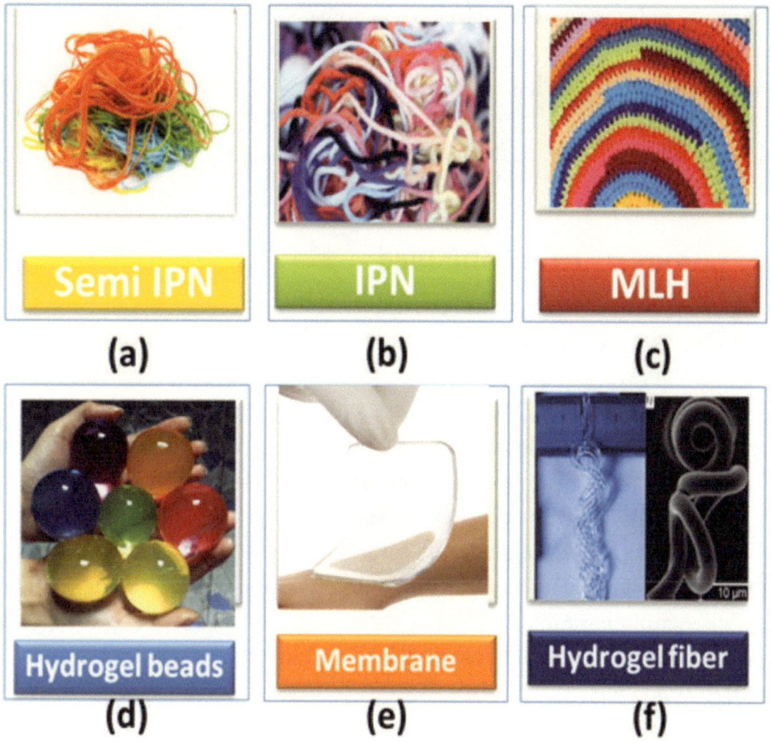

Figure 9. Predominated shape design of hydrogels.

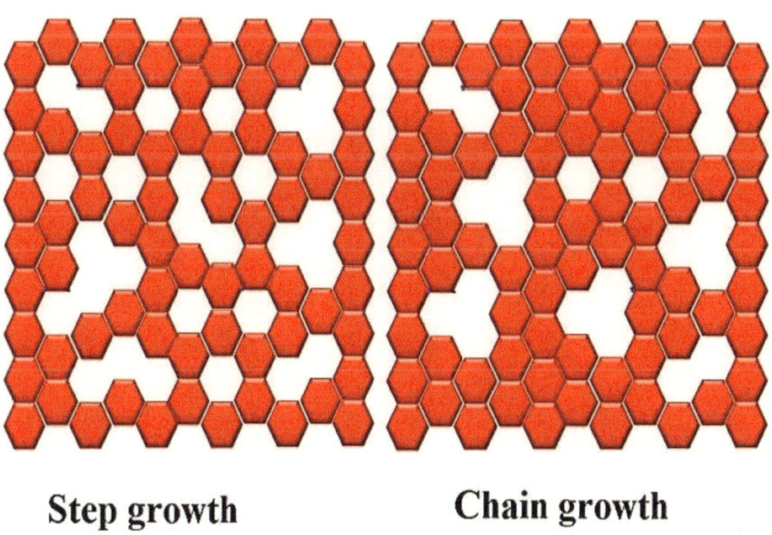

Figure 10. The two routs of polymerization reactions.

of both growths of polymerization reaction. As shown, the polymerization occurs which by chain growth would proceed by random dimerization of the neighboring monomers and then oligomerization formation until the cross-linked network is obtained. Step growth polymerization reaction has formed dimers, trimers, tetramers, etc., until short chains are formed within the combination of them to form cross-linked hydrogel.

6. Types of porous structure obtained

Another attractive feature is the development of new complex hydrogel films with targeted architecture. Porous materials are materials having different pore size structures (from nanometer to millimeter). Hydrogel has a porous structure in size of the micrometer called superporous structure. **Figure 11** shows the pores with different shapes with different accessibility. Almost, the surfaces of pores are hydrophilic, and the void begins to open due to group's repletion of the same charge. The swelling process shows variability of pore structure obtained as (a) closed pores, (b) one side opened pores like cylinder, (c) two sides opened pores like tunnel, (d) one side opened pores like ink bottle shaped, (e) two side opened pores like funnel shaped, (f) pores with rough surface, (g) separated closed pores, (h) interconnected pores, (k) collected pores or density pores, and (p) pores like internal tunnel. In maximum swelling we reach to superporous hydrogel when the pores predominate than the solid network.

Further compression properties of the superporous hydrogel are α-elastin fabricated under 60 bar CO_2 pressure which was comparable with 1 bar. SEM image in **Figure 12** shows the pore size of the hydrogels which was enhanced 20-fold when the pressure was increased from 1 to 60 bar.

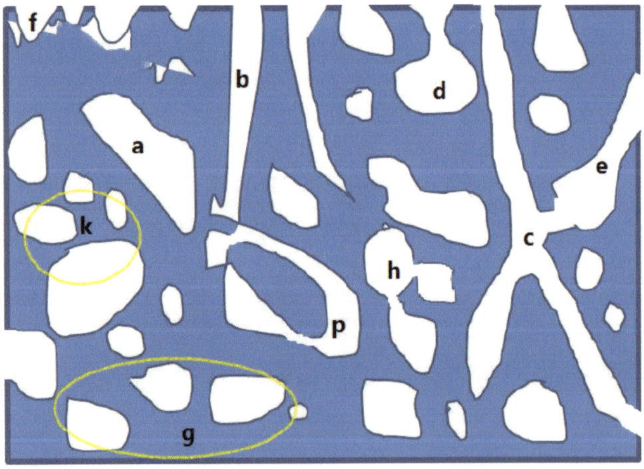

Figure 11. Given the different type of pores where white region is hydrophilic region filled with water and blue region is hydrophobic region or covalent bond of cross-linkage. **Shape**: a, k, g, and h (closed pores: (b) cylindrical open shaped, (c and p) tunnel shaped, (d) ink bottle shaped, (e) funnel shaped, and (f) roughness). **Accessibility**: a, p, h, g, and k (closed pores: c, e, d, f, and p).

http://dx.doi.org/10.5772/intechopen.74698

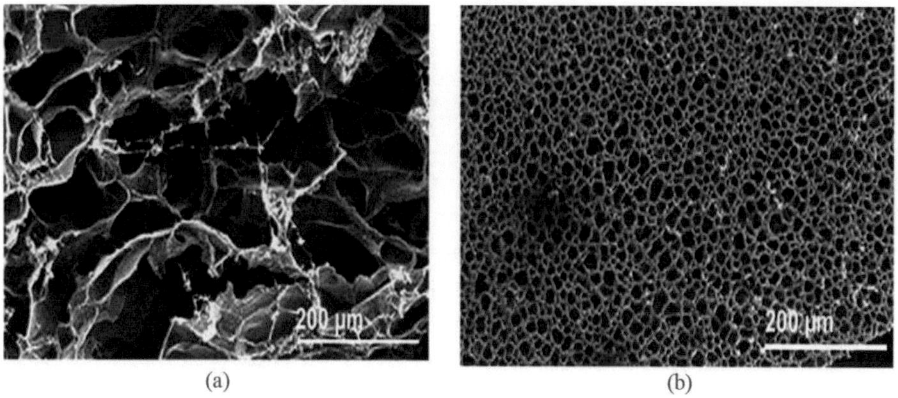

Figure 12. SEM images of (a) the α-elastin hydrogel fabricated at 60 bar CO_2 pressure which was highly porous structure. Comparison of SEM images (b) of α-elastin hydrogel produced under atmospheric CO_2 conditions (1 bar) indicated that a high-pressure CO_2 increased the pore size of the fabricated hydrogels [53].

7. Factors affecting superabsorbent hydrogel

7.1. Density of cross-linking

Increasing the ratio of cross-linked portion leads to slow down the movement of chains, resulting in the decrease in free volume, the pore sizes, and the swelling degree which are also decreasing. This can be observed by SEM analysis or DSC where increased cross-linking causes increase of glassy temperature (T_g) of the polymer [54, 55]. However, in some cases, a decreased cross-linking leads to a decrease of T_g, where nonfreezing (bounded) water molecules are attached to function groups causing a decrease of T_g [56].

7.2. The ratio of hydrophobic/hydrophilic surface area of the hydrogels

The ionization power and number of hydrophilic functional groups along the hydrogel chains and its counterion type play an important role in the degree of swelling. A high proportion of superabsorbent hydrogels are present as acrylates with carboxylic acid functional groups, which in the salt form undergo dissociation upon contact with water. In the dissociated state, the hydrogel network will have a series of functional groups that have the same electric charge and thus repel each other. This leads to expansion of the hydrogel network structure with the further absorption of water molecules. Furthermore, the number of hydrophilic moieties when increasing the swelling could be increased and vice versa.

7.3. Applications of superabsorbent hydrogel

According to the required application, the hydrogels have been tailored and designed to achieve the purpose of applications. The presented section demonstrates the research concerning the characterization of hydrogels on various bases, physical and concoction qualities of these items, and specialized practicality of their usage.

7.4. Internal curing agent for cement

Cross-linking of superabsorbent hydrogels based on poly(acrylamide-co-sodium alginate) by γ-radiation shows higher swelling capacities in basic than in acidic media [57]. This property proposes the use of (PAM-*co*-NaAlg) hydrogel as an internal curing agent for cement. The cement with (PAM-co-NaAlg) hydrogel mixture improves the compressive strength of cement at 0.1 and 0.2 wt%. Intermediate values are found when using 0.1, 0.3, 0.4, and 0.5 wt% of hydrogel. Thus, the maximal improvement percentage on the compressive strength is 0.2% with respect to the hydrogel. This would indicate that using 0.2 wt% of hydrogel is a critical value. Under this value, the hydrogel shows less ability for water retention than what is needed for cement curing. Above this value, the higher amount of hydrogel is contributing to increase voids that cause decreased compressive strength. **Figure 13** demonstrates the procedure for mixing of cement-hydrogel samples in the laboratory for compressive strength test. Mixing of cement and hydrogel samples was carried out with a known w/c ratio at 0.4. The setup of preparation steps is as follows: (1) The known weight of dried PAM-co-NaAlg is added to 100 ml water, and (2) during a specified time period, the hydrogel is allowed to swell, and water uptake is liberated slowly during the hydration of the cement. Then, 40 g of cement (3) is added to the swollen hydrogel, and the slurry of cement is obtained (4). The slurry is stirred well for 3 min and then poured in the aluminum mold (4 × 4 × 4) cm^3 (5). After a period of time (24 hr) which corresponds to the time required obtaining dried cement samples, the mold is removed.

Figure 13. The procedures of preparing cement hydrogel mixture for compressive strength test.

Figure 14a and **b** shows the SEM images of the cement-hydrated (cured) product particles. It is clear that the particles in **Figure 14b** for 0.3 wt% hydrogel are smaller than the particles in **Figure 14a** (without hydrogel). This indicates that the slow absorption of water during cement curing helps the formation of small particles and leads to decreased permeability. Moreover, without hydrogels, the poorly hydrated system is reflecting the compressive strength results.

7.5. Agricultural proposal

Many studies report the use of superabsorbent hydrogel (SAH) in agricultural field (**Figure 15**). Using superabsorbent hydrogel increases soil ability to hold water, so plant growth increases and it can resist drought for a long time [58–60]. Also, adding superabsorbent hydrogel (SAH) to soil improves plants through supplying the plant roots with water, providing soil with potassium ion which is important to retain a K^+/Na^+ homeostasis, and components of polymers held the ions of salt in the drying soil [61]. Superabsorbent hydrogels (PVP/CMC) based on polyvinylpyrrolidone (PVP)/carboxymethyl cellulose (CMC) were prepared by using gamma radiation as an initiator of polymerization reaction [62]. Water and fertilizers are vital factors for producing high-yield agricultural crops. Fertilizers are composed of phosphate (P), potassium (K), and nitrogen (N) nutrients for plants in the form of water-soluble salts which are loaded into (PVP/CMC) hydrogel. The hydrogels show adsorption desorption of the three fertilizers; also, the results revealed that the presence of CMC in the hydrogel improves their water retention capability with high swelling ratio. This indicates that the addition of PVP/CMC hydrogels to soil could improve the water-holding capacity of the soil which has the excellent water absorbing capacity. So, the soil could hold much more water during the irrigation period or raining time than the soil without it and could efficiently reduce irrigation water consumption. It can be concluded that PVP/CMC hydrogel had good water retention capability suggesting their possible use of the prepared superabsorbent hydrogel as a soil conditioner in agriculture applications.

(a) (b)

Figure 14. SEM image of a fracture surface of cement: (a) blank sample without hydrogel and (b) 0.3 wt% of hydrogel.

Figure 15. Used SAH as soil conditioner to support plant growth safely.

7.6. Biomedical proposal

Hydrogels have become very popular due to their unique properties such as high water content, biocompatibility, flexibility, and softness. Natural and synthetic polymers can be a resemblance to the living tissue that opens up several opportunities for applications in biomedical and medicine fields. Currently, hydrogels are used for manufacturing hygiene products, contact lenses, scaffolds, tissue engineering, wound dressings, and drug delivery systems. More developments are expected in drug delivery and tissue engineering. A hydrogel based on poly-2-hydroxyethylmethacrylate (PHEMA) as a synthetic biocompatible material is used as contact lens applications [63, 64]. Wound dressing is an effective hydrogel dressing that relies on an understanding of the healing process. Healing can be hindered by many factors such as infection, abnormal bacterial presence or desiccation, maceration, necrosis, pressure, edema, and trauma [65]. The "ideal" wound management product should absorb excess toxins and exudate, keep a good moisture between the wound and the dressing with increasing collagen production, preserve the wound from external sources of infection, prevent excess heat at the wound, have good permeability to gases, be supplied completely sterile, and be easy to remove without further trauma to the wound [66]. Their high water content allows vapor and oxygen transmission to the wounds such as pressure sores, leg ulcers, surgical and necrotic wounds, lacerations, and burns. They seem to play an important role as emergency burn treatment alone or in combination with other products, thanks to their cooling and hydrating effect [67]. Hydrogels have attracted noticeable interest for their use in drug delivery due to their unique physical properties [68]. The high porosity that characterizes SAH can easily absorb and desorb drugs easily by adjusting the density of cross-links in their matrix and the affinity to water.

Tissue engineering is the application of the principles and methods of engineering and life sciences toward fundamental understanding of the structure–function relationship in normal and pathological mammalian tissues and the development of biological substitutes for

http://dx.doi.org/10.5772/intechopen.74698

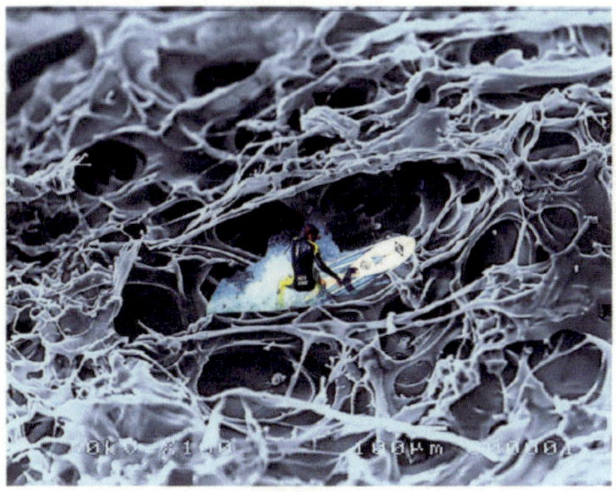

(a)

(b)

(c)

Figure 16. SEM image of hydrogel in funny situation.

the repair or regeneration of tissue or organ function [69]. Tissue engineering is a more recent application of hydrogels, in which they can be applied as space-filling agents, as delivery vehicles for bioactive substances or as three-dimensional structures that organize cells and present stimuli to ensure the development of a required tissue. Space-filling agents are the most commonly used group of scaffolds, and they are employed for bulking, to prevent adhesion, and as biological "glue." Drugs can be delivered from hydrogel scaffolds in numerous applications including promotion of angiogenesis and encapsulation of secretory cells. Additionally, hydrogel scaffolds have also been applied to transplant cells and to engineer many tissues in the body, including the cartilage, bone, and smooth muscle [11].

7.6.1. Environmental proposal

Hydrogel is an eco-friendly material that has many uses as water purification and air purification. Hydrogel as a new type of adsorbent for water purification is a composite with graphene oxide [70]. These materials usually exhibit high-capacity adsorption toward water pollutants and air pollutants [71]. Due to their highly developed porous structure, hydrogel fabricated with other materials such as graphene oxide and zeolite exhibits a high capacity for gas adsorption and selectivity for gas separation. Researches are still on developing the efficiency of hydrogel gas adsorbents with a good stability, recyclability, and substantial capacity. Because of their porous structure, and their environment stability, hydrogel is a good candidate for gas adsorption.

7.6.2. Gallery of hydrogel

Figure 16 shows funny imaginary SEM pictures of swollen hydrogel (a) PVP/PAAc, (b) PVP/PAAc after swelled in 1 M KCl, and (c) CMC/PVP loaded with three kinds of salts.

Author details

Mohamed Mohamady Ghobashy

Address all correspondence to: moahmed.ghobashy@eaea.org.eg

Hydrogel Lab, Radiation Research of Polymer Department, National Center for Radiation Research and Technology (NCRRT), Atomic Energy Authority, Nasr City, Cairo, Egypt

References

[1] Ratner BD, Hoffman AS. Hydrogels for Medical and Related Applications. Andrade JD, editor. Washington, D.C.: American Chemical Society; 1976. pp. 1-36

[2] Park K, Shalaby WSW, Park H. Biodegradable Hydrogels for Drug Delivery. Basel: Technomic Publishing Company, Inc; 1993. pp. 1-12

[3] Anderson JM. Polymeric Biomaterials. Piskin E, Hoffman AS, editors. Boston, MA: Martinus Nijhoff Publishers; 1986. pp. 29-39

[4] Aalaie J, Rahmatpour A, Vasheghani-Farahani E. Rheological and swelling behavior of semi-interpenetrating networks of polyacrylamide and scleroglucan. Polymers for Advanced Technologies. 2009;**20**:1102-1106

[5] Frisch HL. Sorption and transport in glassy polymers—A review. Polymer Engineering & Science. January 1980;**20**(1):2-13

[6] Ganji F, Vasheghani-Farahani S, Vasheghani-Farahani E. Theoretical description of hydrogel swelling: A review. Iranian Polymer Journal. 2010;**19**(5):375-398

[7] Ahmed EM. Hydrogel: Preparation, characterization, and applications: A review. Journal of Advanced Research. 2015;**6**(2):105-121

[8] Kim IY, Seo SJ, Moon HS, et al. Chitosan and its derivatives for tissue engineering applications. Biotechnology Advances. 2008;**26**(1):1-21

[9] Liang Y, Liu W, Han B, et al. An in situ formed biodegradable hydrogel for reconstruction of the corneal endothelium. Colloids Surface B. 2011;**82**(1):1-7

[10] Davidenko N, Campbell JJ, Thian ES, Watson CJ, Cameron RE. Collagen-hyaluronic acid scaffolds for adipose tissue engineering. Acta Biomaterialia. 2010;**6**(10):3957-3968

[11] Drury JL, Mooney DJ. Hydrogels for tissue engineering: Scaffold design variables and applications. Biomaterials. 2003;**24**(24):4337-4351

[12] Liu SQ, Tay R, Khan M, Ee PLR, Hedrick JL, Yang YY. Synthetic hydrogels for controlled stem cell differentiation. Soft Matter. 2010;**6**(1):67-81

[13] Hunt NC, Grover LM. Cell encapsulation using biopolymer gels for regenerative medicine. Biotechnology Letters. 2010;**32**(6):733-742

[14] Lee J, Cuddihy MJ, Kotov NA. Three-dimensional cell culture matrices: State of the art. Tissue Engineering Part B. 2008;**14**(1):61-86

[15] Zhu J. Bioactive modification of poly(ethylene glycol) hydrogels for tissue engineering. Biomaterials. 2010;**31**(17):4639-4656

[16] Geckil H, Xu F, Zhang XH, Moon S, Demirici U. Engineering hydrogels as extracellular matrix mimics. Nanomedicine. 2010;**5**(3):469-484

[17] Tan H, Wu J, Lao L, Gao C. Gelatin/chitosan/hyaluronan scaffold integrated with PLGA microspheres for cartilage tissue engineering. Acta Biomaterialia. 2009;**5**(1):328-337

[18] Rosellini E, Cristallini C, Barbani N, Vozzi G, Giusti P. Preparation and characterization of alginate/gelatin blend films for cardiac tissue engineering. Journal of Biomedical Materials Research. Part A. 2009;**91**(2):447-453

[19] Liu Y, Chan-Park MB. Hydrogel based on interpenetrating polymer networks of dextran and gelatin for vascular tissue engineering. Biomaterials. 2009;**30**(2):196-207

[20] Lin CC, Metters AT. Hydrogels in controlled release formulations: Network design and mathematical modeling. Advanced Drug Delivery Reviews. 2006;**58**(12-13):1379-1408

[21] Hejcl A, Sedy J, Kapcalova M, et al. HPMA-RGD hydrogels seeded with mesenchymal stem cells improve functional outcome in chronic spinal cord injury. Stem Cells Development. 2010;**19**(10):1535-1546

[22] Zhu J, Marchant RE. Design properties of hydrogel tissue-engineering scaffolds. Expert Review of Medical Devices. 2011;**8**(5):607-626. DOI: 10.1586/erd.11.27

[23] Jao D, Mou X, Hu X. Tissue Regeneration: A Silk Road. Puoci F, ed. Journal of Functional Biomaterials. 2016;**7**(3):22. doi:10.3390/jfb7030022

[24] Ghobashy MM, Elhady MA. pH-Sensitive wax emulsion copolymerization with acrylamide hydrogel using gamma irradiation for dye removal. Radiation Physics and Chemistry. May 2017;**134**:pp. 47-55

[25] Kulkarni RV, Sa B. Evaluation of pH-sensitivity and drug release characteristics of (polyacrylamide-grafted-xanthan)-carboxymethyl cellulose-based pH-sensitive interpenetrating network hydrogel beads. Drug Development and Industrial Pharmacy. 2008; **34**(12):1406-1414

[26] Fong H, Reneker DH. Elastomeric nanofibers of styrene-butadiene-styrene triblock copolymer. Journal of Polymer Science Part B Polymer Physics. 1999;**37**(24):3488-3493

[27] Osman H, Ismail H, Mariatti M. Polypropylene/natural rubber composites filled with recycled newspaper: Effect of chemical treatment using maleic anhydride-grafted polypropylene and 3-aminopropyltriethoxysilane. Polymer Composites. 2012;**33**(4):609-618

[28] Ghobashy MM, Awad A, Elhady MA, Elbarbary AM. Silver rubber-hydrogel nanocomposite as pH-sensitive prepared by gamma radiation: Part I. Cogent Chemistry. 2017; **3**(1):1328770

[29] Tabata Y. Biomaterial technology for tissue engineering applications. Journal of Royal Society Interface. 2009;**6**:S311-S324

[30] Yokoyama F, Masada I, Shimamura K, Ikawa T, Monobe K. Morphology and structure of highly elastic poly(vinyl alcohol) hydrogel prepared by repeated freezing-and-melting. Colloid & Polymer Science. 1986;**264**:595-601

[31] Gupta S, Goswami S, Sinha A. A combined effect of freeze--thaw cycles and polymer concentration on the structure and mechanical properties of transparent PVA gels. Biomedical Materials. 2012;**7**(1):015006

[32] Kim TH, An DB, Oh SH, Kang MK, Song HH, Lee JH. Creating stiffness gradient polyvinyl alcohol hydrogel using a simple gradual freezing-thawing method to investigate stem cell differentiation behaviors. Biomaterials. 2015;**40**:51e60

[33] Hoffman AS. Hydrogels for biomedical applications. Advanced Drug Delivery Reviews. 2012;**64**:18-23

[34] Cascone MG, Lazzeri L, Sparvoli E, Scatena M, Serino LP, Danti S. Morphological evaluation of bioartificial hydrogels as potential tissue engineering scaffolds. Journal of Materials Science: Materials in Medicine. 2004;**15**(12):1309-1313

[35] ZHANG Y, HUI B, YE L. Reactive toughening of polyvinyl alcohol hydrogel and its wastewater treatment performance by immobilization of microorganisms. RSC Advances. 2015;**5**(111):91414-91422

[36] Bitar KN, Zakhem E. Design strategies of biodegradable scaffolds for tissue regeneration. Biomedical Engineering and Computational Biology. 2014;**6**:13

[37] Fernandes LL, Resende CX, Tavares DS, Soares GA, Castro LO, Granjeiro JM. Cytocompatibility of chitosan and collagen-chitosan scaffolds for tissue engineering. Polímeros. 2011;**21**(1):1-6 Epub February 11, 2011.https://dx.doi.org/10.1590/S0104-14282011005000008

[38] Mitra T, Sailakshmi G, Gnanamani A, Mandal AB. Studies on cross-linking of succinic acid with chitosan/collagen. Materials Research. 2013;**16**(4):755-765 Epub April 23, 2013. https://dx.doi.org/10.1590/S1516-14392013005000059

[39] MITRA T, SAILAKSHMI G, GNANAMANI A. Could glutaric acid (GA) replace glutaraldehyde in the preparation of biocompatible biopolymers with high mechanical and thermal properties? Journal of Chemical Sciences. 2014;**126**(1):127-140

[40] Ghobashy MM, Khafaga MR. Radiation synthesis and magnetic property investigations of the graft copolymer poly (ethylene-g-acrylic acid)/Fe_3O_4 film. Journal of Superconductivity and Novel Magnetism. 2016;**30**(2):401-406

[41] Ghobashy MM, Abdel Reheem AM, Mazied NA. Ion etching induced surface patterns of blend polymer (poly ethylene glycol – poly methyl methacrylate) irradiated with gamma rays. International Polymer Processing. 2017;**32**(2):174-182

[42] Ghobashy MM, Abdeen ZI. Influence of gamma irradiation on the change of the characterization of elastomeric polyurethane. Advanced Science, Engineering and Medicine. 2016;**8**(9):736-739

[43] Ghobashy MM, Khozemey E. Sulfonated gamma-irradiated blend poly(styrene/ethylene-vinyl acetate) membrane and their electrical properties. Advances in Polymer Technology. 2016. DOI: 10.1002/adv.21781

[44] Ghobashy MM. Effect of sulfonated groups on the proton and methanol transport behavior of irradiated PS/PEVA membrane. International Journal of Plastics Technology:pp. 1-14

[45] An JC. Synthesis of the combined inter-and intra-crosslinked nanohydrogels by e-beam ionizing radiation. Journal of Industrial and Engineering Chemistry. 2010;**16**(5):657-661

[46] Jenkins AD, Kratochvil RF, Stepto RFT, Suter UW. Glossary of basic terms in polymer science. Pure and Applied Chemistry. 1996;**68**:2287-2311

[47] Higgins W, Kozlovskaya V, Alford A, Ankner J, Kharlampieva E. Stratified temperature-responsive multilayer hydrogels of poly (N-vinylpyrrolidone) and poly (N-vinylcaprolactam): Effect of hydrogel architecture on properties. Macromolecules. 2016;**49**(18):6953-6964

[48] Liu X-j, Zhang Y-m, Li X-s. Tough biopolymer IPN hydrogel fibers by Bienzymatic crosslinking approach. Chinese Journal of Polymer Science. 2015;**33**(12):1741-1749. DOI: 10.1007/s10118-015-1717-9

[49] Cheng Z, Cui M, Shi Y, Qin Y, Zhao X. Fabrication of cell-laden hydrogel fibers with controllable diameters. Micromachines. 2017;**8**(5):161

[50] Gaharwar AK, Schexnailder PJ, Dundigalla A, White JD, Matos-Pérez CR, Cloud JL, Schmidt G, et al. Highly Extensible Bio-Nanocomposite Fibers. Macromolecular Rapid Communications. 2011;**32**(1):50-57

[51] Jiang K, Liu Y, Yan Y, Wang S, Liu L, Yang W. Combined chain-and step-growth dispersion polymerization toward PSt particles with soft, clickable patches. Polymer Chemistry. 2017;**8**(8):1404-1416

[52] Shih H, Lin C-C. Crosslinking and degradation of step-growth hydrogels formed by Thiol-Ene photo-click chemistry. Biomacromolecules. 2012;**13**(7):2003-2012 PMC. Web. 21 Aug. 2017

[53] Annabi N, Mithieux SM, Boughton EA, Ruys AJ, Weiss AS, Dehghani F. Synthesis of highly porous crosslinked elastin hydrogels and their interaction with fibroblasts in vitro. Biomaterials. 2009;**30**(27):4550-4557

[54] Tsukasa M, Yuuki T, Sachiko A, Takahiko I, Akie H, Keiko E. Role of boric acid for a poly (vinyl alcohol) film as a cross-linking agent: Melting behaviors of the films with boric acid. Polymer. 29 October 2010;**51**(23):5539-5549

[55] Simon SL, Gillham JK. Reaction kinetics and TTT cure diagrams for off-stoichiometric ratios of a high-Tg epoxy/amine system. Journal of Applied Polymer Science. 1992; **46**:1245-1270

[56] Sanyang ML, Sapuan SM, Jawaid M, Ishak MR, Sahari J. Effect of plasticizer type and concentration on tensile, thermal and barrier properties of biodegradable films based on sugar palm (*Arenga pinnata*) starch. Polymers. 2015;**7**:1106-1124

[57] Ghobashy MM, Bassioni G. pH stimuli-responsive poly(acrylamide-co-sodium alginate) hydrogels prepared by γ-radiation for improved compressive strength of concrete. Advances in Polymer Technology. 2017;**00**:1-11 https://doi.org/10.1002/adv.21870

[58] Khalilpour A. Study the application of superabsorbent polymer (BT773) on controlling soil erosion and conservation. Report of Research Project. Tehran Research Center of Natural Resources. Ministry of Jihad Agriculture. Tehran. Iran; 2001

[59] DeVarennes AD, Queda C. Application of an insoluble polyacrylate polymer to copper-contaminated soil enhances plant growth and soil quality. Soil Use and Management. 2005;**21**:410-414

[60] Shooshtarian S, Abedi-Kupai J, TehraniFar A. Evaluation of application of superabsorbent polymers in green space of arid and semi-arid regions with emphasis on Iran. International Journal of Forest, Soil and Erosion. 2012;**2**(1):24-36

[61] Ataei H, Ghorbani M. Application of superabsorbent hydrogel in green space. Journal of Automotic Urban Services. 2001;**36**:42-45

http://dx.doi.org/10.5772/intechopen.74698

[62] Ahmed EM, Mohamed MG. Controlled release fertilizers using superabsorbent hydrogel prepared by gamma radiation. Radiochimica Acta 105.10. 2017:865-876

[63] Wichterle O, Lim D. Hydrophilic gels for biological use. Nature. 1960;**185**(4706):117-118

[64] Maldonado-Codina C, Efron N. Optometry. Practice. 2003;**4**:101-115

[65] Turner TD. Pharmaceutical Journal. 1979;**222**:421-424

[66] Jones V, Grey JE, Harding KG. BMJ. 2006;**332**:777-780

[67] Osti E, Osti Ann F. Burns Fire Disasters. 2004;**3**:137-141

[68] Hoare TR, Kohane DS. Polymer. 2008;**49**:1993-2007

[69] Chapekar MS. Journal of Biomedical Materials Research. 2000;**53**:617-620

[70] Chen Y, Chen L, Bai H, Li L. Graphene oxide–chitosan composite hydrogels as broad-spectrum adsorbents for water purification. Journal of Materials Chemistry A. 2013;**1**(6):1992-2001

[71] Xiong X et al. Preparation functionalized graphene aerogels as air cleaner filter. Procedia Engineering. 2015;**121**:957-960

Chapter 4

Hydrogels Applied for Conformance-Improvement Treatment of Oil Reservoirs

Fernanda G. C. Tessarolli, Ailton S. Gomes and Claudia R. E. Mansur

Additional information is available at the end of the chapter

http://dx.doi.org/10.5772/intechopen.73204

Abstract

This chapter aims at presenting a review of gelling polymer systems that are commercially available or under academic development with potential to control the anisotropic permeability profile of heterogeneous oil reservoirs. In these reservoirs, the oil recovery and sweep efficiency tend to be low, even after applying secondary and enhanced oil recovery methods, because the injected fluid flows preferably through the matrix's most permeable regions leaving behind part of the displaceable oil retained at the nonswept volume. For that, cross-linked polymers can be used to plug the high-permeability main paths by means of: (i) the formation of an *in situ* hydrogel or (ii) the adsorption or swelling of pre-cross-linked hydrogel within the reservoir pores, thus causing the diversion of the subsequently injected fluid to low-permeability zones and/or preventing the channeling and early breakthrough of the injected fluid (water or gas) in production wells. The selection of the most suitable hydrogel for the reservoir conformance-improvement treatment should take into account the nature of the conformance problem, the reservoir's lithology, mineralogy, temperature, pH value, salinity, and hardness of the formation water, as well as the gelling system toxicity and cost.

Keywords: hydrogel, polymer, conformance improvement, reservoir, oil and gas

1. Introduction

During the productive life cycle of an oil reservoir, primary, secondary, and enhanced oil recovery methods can be applied to improve the overall hydrocarbon recovery.

Initially, during primary recovery, the oil production is accomplished by the use of the natural reservoir energy as well as artificial lift and well stimulation methods that do not directly affect the driving force of the porous medium. Nevertheless, the reservoir primary energy is progressively dissipated due to the decompression of fluids and to forces (viscous and capillary forces) imposed on the matrix flow.

In order to minimize the negative impacts of primary energy dissipation and increase oil recovery, two strategies may be used:

- addition of an artificial secondary energy to the reservoir by means of the injection of fluids, and/or
- reduction of the viscous and/or capillary forces acting on the reservoir.

The injection of fluids into the formation (water or immiscible gas) to displace hydrocarbons from the pores of the reservoir matrix with no chemical or thermodynamic interactions between the injected fluid and oil/reservoir rock is known as secondary recovery.

Together, primary and secondary oil recovery methods produce around 33% of a reservoir's initial-oil-in-place (OOIP) [1]. This low oil recovery inherent in these methods occurs due to: (i) high interfacial tension between oil and injected fluids and/or (ii) high viscous oil in the reservoir[1].

In these cases, the use of enhanced oil recovery (EOR) methods, also known as improved oil recovery (IOR) methods, is recommended, as they act on the sweep efficiency and/or on the displacement efficiency of injected fluids.

However, in fractured and/or stratified reservoirs with anisotropic permeability profiles (conformance problems), the sweep and displacement efficiencies tend to be low even after applying secondary and enhanced oil recovery methods.

In these heterogeneous porous media, the injected fluid tends to flow preferably through high-permeability zones and/or anomalies[2] (preferential paths), failing to recover part of the displaceable oil in low-permeability-unswept zones of the reservoir.

The channeling of the injected fluid through these high-permeability zones and/or anomalies, besides reducing the total oil recovery, can also be responsible for the early breakthrough of the injected fluid in production wells, thus increasing the operational costs associated with the separation, treatment, and disposal of the produced fluid (e.g. water).

Therefore, to remedy these problems, several authors [2–6] have proposed the injection of gelling polymer systems[3] to selectively flow through the high-permeability zones or anomalies, temporarily plugging them with a barrier (hydrogel) (**Figure 1**) in order to:

[1]When there is high interfacial tension between the injected fluid and the displaced fluid, the injected fluid capacity to displace oil from the reservoir is significantly reduced, inducing bypass of residual oil by the injected fluid. When the viscosity of the injected fluid is much lower than that of the fluid being displaced, the former displaces much more easily in the porous medium, finding preferential paths and moving quickly toward the producing wells. As the injected fluid does not spread properly inside the reservoir, the oil is trapped in large bulks of reservoir rock in which displacement does not occur.

[2]Fracture networks (both natural and hydraulically induced), faults, interconnected vugular porosity, caverns, and localized matrix reservoir rock with permeabilities greater than 2 Da.

[3]The term gelling system or gelant refers to a polymer + cross-linker solution or a microgel dispersion before any appreciable cross-linker has occurred. The term gel is used when the gelling system has attained either partial or full crosslinking maturation.

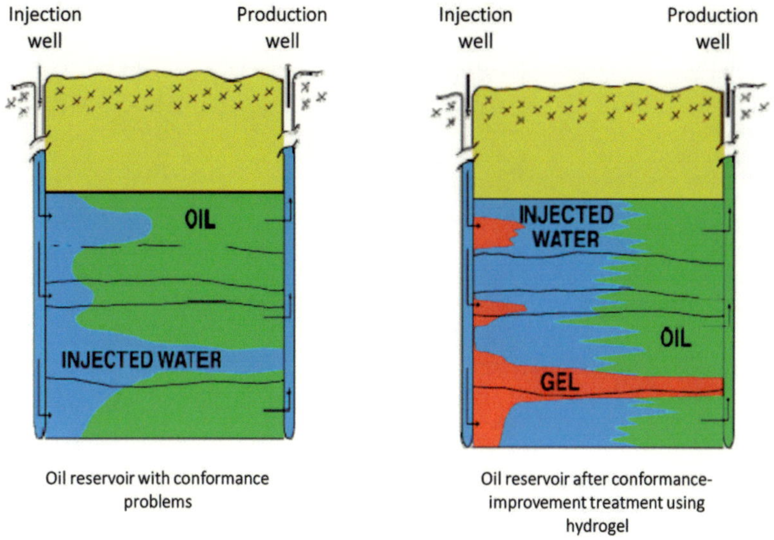

Figure 1. Schematic of an oil reservoir conformance-improvement treatment with hydrogel.

- divert the injected fluid flow from high permeability, low-oil-saturation reservoir flow paths to low-permeability, high-oil-saturation flow paths (in-depth profile control), improving flood sweep efficiency, and producing incremental oil, or
- block high-permeability zones or anomalies located near wellbore modifying the injection profile and/or preventing the channeling and early breakthrough of the injected fluid (water or gas) in production wells (water or gas shut off), thus reducing overall oil-production operational costs.

This chapter aims at presenting a review of gelling polymer systems commercially available or under academic development, with potential to control the anisotropic permeability profile of heterogeneous oil reservoirs.

2. Gelling polymer systems for conformance control of oil reservoirs

Hydrogels used to control the anisotropic permeability profile of oil reservoirs are cross-linked polymers, swellable in water that retain the solvent within their three-dimensional structures without dissolving them [7–9].

Polyacrylamide homopolymer (PAM) and acrylamide copolymers such as: partially hydrolyzed polyacrylamide (PHPA), copoly(acrylamide-t-butyl acrylate) (PAtBA), copoly(acrylamide-2-acrylamido-2-methyl-propanesulfonic acid) (PAM-AMPS), and copoly(acrylamide-N-vinyl-2-pyrrolidone) (PAM-NVP)[4] are the most widely used polymers for conformance-improvement

[4]The incorporation of AMPS and/or NVP groups in acrylamide-copolymer chains prevents the acrylamide groups from autohydrolyzing at high temperatures, reducing the polymer susceptibility to precipitate out of the solution in the presence of hardness divalent ions (i.e. Ca^{2+} or Mg^{2+}). For this reason, AMPS-NVP-acrylamide copolymers are mainly applied for the conformance control of reservoirs with harsh conditions (i.e. temperature > 90°C and salinity >100,000 ppm TDS) [10, 11].

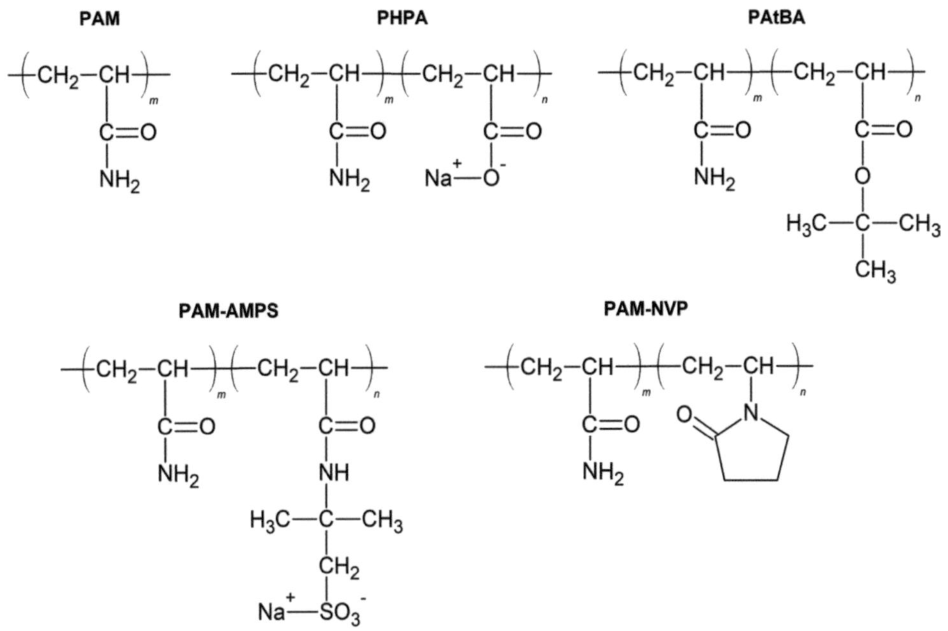

Figure 2. Acrylamide-based polymers used for conformance control of oil reservoirs.

treatments of oil reservoirs **Figure 2**. Biopolymers such as xanthan gum, guar gum, chitosan, starch, cellulose, and scleroglucan have also been studied [7, 11, 12].

These polymers can be cross-linked by organic and inorganic cross-linkers forming chemical hydrogels through covalent bonding or physical hydrogels through ionic complexing, hydrogen bonding, polymer entanglements, van der Waals interactions, and/or hydrophilic interactions [13, 14].

The formed hydrogel structure and properties depend on the gelling system components (polymer, cross-linker, and solvent), the concentration of reagents, and the reaction conditions (pH value, salinity, and temperature). An increase in the polymer and/or cross-linker concentration in the gelling system results in a hydrogel with a rigid structure due to a larger number of cross-links per unit of chain length [13].

For conformance-improvement treatment of heterogeneous oil reservoirs, the gelling system must [11, 12, 15]:

- be formulated with low concentrations of relatively inexpensive environmentally acceptable and friendly chemicals;
- have good injectivity and propagation within the matrix reservoir rock;
- have controllable and predictable gelation time;
- have thermal, mechanical, and biological stability under reservoir conditions; and
- provide a broad range of gel strengths, including rigid gels.

http://dx.doi.org/10.5772/intechopen.73204

Groups	Gelling polymer systems	Reservoir temperature	Formation water salinity
In situ cross-linked polymer systems	Gelling systems cross-linked with metal ions [11, 20–35]	up to 90°C	up to 30,000 mg/L TDS
	Gelling systems cross-linked with organic compounds [19, 42–57]	30–180°C	30,000–100,000 mg/LTDS
	Gelling systems with no cross-linker [58, 59]	up to 90°C	up to 30,000 mg/L TDS
Pre–cross-linked polymer systems	Colloidal dispersion gels [61–63]	40–90°C	up to 15,000 mg/L TDS
	Microgel systems [4, 64–68]	up to 90°C	insensitive to salinity variations
	BrightWater™ systems [69–73]	35–140°C	up to 120,000 mg/L TDS
	Preformed particle gels [74–76]	up to 120°C	up to 300,000 mg/L TDS
	pH-sensitive gel systems [60, 77–79]	up to 90°C	up to 30,000 mg/L TDS

Table 1. Summary of gelling polymer systems used for conformance control of oil reservoirs.

Furthermore, the porous matrix should present: (i) zones with high-permeability contrasts (e.g. 10:1), (ii) high thickness ratios (e.g. less-permeable zones being 10 times thicker than high-permeability zones, and (iii) relatively low oil viscosities [16].

Moreover, hydrogels that suffer severe dehydration and syneresis, greatly reducing their volume over time under reservoir conditions, are not suitable for reservoir permeability profile treatments[5].

Several gelling systems have been developed (**Table 1**) for the treatment of high-permeability zones of a broad range of reservoir conditions[6] by means of the:

- formation of an *in situ* hydrogel (Section 2.1); or
- adsorption/swelling of pre–cross-linked hydrogel within the reservoir rock (Section 2.2).

2.1. *In situ* cross-linked polymer systems

In *in situ* cross-linked polymer systems, a polymer solution and a cross-linking agent are injected together into the porous matrix forming an *in situ* hydrogel within the high-permeability zones of the reservoir [19–22].

[5]Gel dehydration can occur in fractures subjected to high pressure gradients. Gel syneresis can result from the increase in the cross-linking density due to excessive cross-linker agent addition and/or excessive cross-linking sites developed on the polymer chain overtime (e.g. autohydrolysis of acrylamide polymers at high temperatures) [17].

[6]Typical conditions found in oil reservoirs are: temperature ranging from 50 to 150°C, pressure between 10 and 50 MPa, salinity from 2000 up to 300,000 mg/L of Total Dissolved Solids (TDS) (with 1 to 35 mg/L of divalent ions), pH value between 4 and 8, and permeability from 10 to 450 mD [18].

Several cases of success of field applications of *in situ* cross-linked gelling systems have been reported for the conformance-improvement treatment of high-permeability zones located near-wellbore (with radial penetration into the matrix lower than 5 m) and far-wellbore (with radial penetration into the matrix greater than 5 m). However, *in situ* cross-linked gelling systems still face a number of operational difficulties regarding the control of the gelation kinetics, the efficient mixing of the polymer and cross-linker within the reservoir, the prevention of undesirable separation of the gelant components into the heterogeneous matrix, and the risk of plugging the whole formation or areas containing oil [23].

The major *in situ* cross-linked polymer systems commercially available or under academic development are presented below.

2.1.1. Gelling systems cross-linked with metal ions

A wide variety of metal ions (e.g. chromium (III), aluminum (III), zirconium (IV), and titanium (III)) can be used to cross-link water-soluble polymers with carboxylates groups such as: polyacrylamide copolymers (e.g. PHPA, PAtBA, and PAM-AMPS) and biopolymers (e.g. xanthan gum, starch, guar, carboxymethylcellulose, scleroglucan, and lignosulfonates) [11, 20].

These gelling systems form physical bulk hydrogels with continuous semi-solid three-dimensional structures by means of ionic interactions between the multivalent metal cations of the cross-linker and the carboxylate/anions of the polymer chains. The cross-linking rates of these systems can be controlled by varying the polymer and cross-linker concentration, the polymer hydrolysis degree, the solution pH value, and by adding cross-linker's retardants (e.g. acetates, propionates, malonates, and ascorbates) [11, 24–26].

PHPA-chromium (III) ions-based formulations are one of the most commonly used systems applied for the permeability profile treatment of oil reservoirs. These systems have been studied extensively as well as tested in laboratory and field for in-depth conformance control of reservoirs with temperatures up to 90°C [27, 28].

Routson and Caldwell [29] were the first to observe that the injection of a dilute solution of chromium hydroxide and PHPA into a porous medium formed a gel that reduced the matrix permeability. Several authors have proposed changes in the formulation developed by Routson and Caldwell [29] in order to reduce the gelant environmental impact, to increase the gelation time, as well as to improve the formed hydrogel strength and stability such as:

- the use of alternative metal ions with lower toxicity when compared to chromium (III) ions, for example, aluminum (III), zirconium (IV), and titanium (III) ions [11, 30–32];
- the use of metal ions retardants to increase the gelant gelation time, for example, acetates, propionates, malonates, and ascorbates [17, 32–35]. Malonate ions are 33 times slower than acetate ions in the gelation of PAM-AMPS at 120°C. Under the same conditions, ascorbate ions are 51 times slower than acetate ions. Nevertheless, the use of complexes of metal ions results in the formation of weak hydrogels [34];

http://dx.doi.org/10.5772/intechopen.73204

- the encapsulation of the cross-linker by polyelectrolyte nanogels of polyethylenimine (PEI) and dextran sulfate (DS) to control the release of the metal ions into the porous matrix, forming hydrogels with gelation time from 5 to 30 days, depending on the reservoir temperature [36, 37];
- the use of biopolymers less shear-sensitive than the polyacrylamide in chromium (III)-cross-linked formulations to reduce the polymer degradation when subjected to prolonged shear into the reservoir, for example, xanthan gum, scleroglucan, *Alcaligenes* bacteria-produced biopolymer and lignosulfonate by-product from paper industry [38, 39]; and
- the addition of nanoclays (e.g. kaolinite, laponite, and montmorillonite) to the PHPA/PAM-AMPS-chromium (III) acetate formulations forming nanocomposite hydrogels with good mechanical strength (modulus, elasticity, and deformability) and improved syneresis resistance [40, 41].

2.1.2. Gelling systems cross-linked with organic compounds

Various organic compounds (e.g. phenol, resorcinol and formaldehyde, hydroquinone, hexamethylenetetramine, chitosan, modified starch and polyethylenimine) can be used to cross-link water-soluble acrylamide-based polymers (e.g. PAM, PHPA, PAtBA, PAM-AMPS, and PAM-NVP) and biopolymers (e.g. modified starch and chitosan) [11, 20, 42–46].

In these systems, a chemical bulk hydrogel with continuous semi-solid three-dimensional network is formed by the reaction of the cross-linker functional groups with the polymer chain functional groups through covalent bonding. These systems have been extensively studied and tested in laboratory and field, especially for harsh reservoir environments (high temperature, salinity, and/or pH value). They have higher gelation times and form hydrogels with higher thermal stability when compared to metal ion–cross-linked systems and are used for the conformance-improvement treatment of reservoirs with temperatures around 30 up to 180°C [43, 47–49].

Polyacrylamide cross-linked with the product of phenol-formaldehyde reaction formed a hydrogel that remained stable for 13 years at 121°C in a laboratory aging test. This gelling system was successfully applied for the treatment of oil reservoir conformance problems [7, 50].

Despite the good thermal stability shown under harsh conditions, phenol-formaldehyde cross-linked gelling systems have great disadvantages, the phenol toxicity and the formaldehyde carcinogenicity. Thus, less toxic and more environmentally friendly organic cross-linking agents have been studied.

Ortho and para-aminobenzoic acids, m-aminophenol, phenyl acetate, phenyl salicylate, salicylamide, salicylic acid, and furfuryl alcohol were identified as substitutes for phenol and hexamethylenetetramine (HMTA) proved to be suitable to replace formaldehyde, forming stable hydrogels with polyacrylamide [51].

Gelling systems based on polyacrylamide cross-linked with hydroquinone (HQ), hexamethylenetetramine (HMTA), and sodium bicarbonate-formed hydrogels that remained stable for 12 months, at 149°C, and for 5 months, at 176.7°C [52, 53].

The performance of gelling systems based on acrylamide copolymers and biopolymers (starch and chitosan) cross-linked with chitosan, starch, and polyethylenimine (PEI) has been evaluated for the conformance control of reservoir with temperatures above 80°C. Laboratory tests showed that PAtBA cross-linked with PEI formed a stable hydrogel at 156°C for 3 months. Furthermore, acrylamide-based polymers-PEI systems were successfully applied to treat conformance problems of carbonate and sandstone reservoirs with temperatures around 130 and 80°C, respectively [47, 49, 54–57].

2.1.3. Gelling systems with no cross-linker

Kansas University Super Polymer One (KUSP1) is a gelling system developed by the University of Kansas based on a biopolymer–polysaccharide (β-1,3-polyglucane)—produced via fermentation by *Alcaligenes faecalis* and *Agrobacterium* bacteria.

This nontoxic biopolymer gels in the absence of cross-linker when the pH value is reduced below 10.8. This gelation process is reversible so that the formed hydrogel can be dissolved by increasing the pH of the solution and can be re-gelled by reducing it.

Field tests were carried out with KUSP1 gelling system to control the permeability profile of oil reservoirs [58, 59].

2.2. Pre–cross-linked polymer systems

In pre–cross-linked polymer systems, the polymer chains are cross-linked in the surface facilities prior to be injected into the reservoir, or at least, are partially gelled in the wellbore. These systems are injected into the reservoir in the form of dispersed aggregates or particles that swell in water within the matrix plugging thief zones with high permeability.

These dispersed aggregates or preformed particles do not have the same drawbacks faced by the *in situ* cross-linked systems, such as: reaction control problems, change of gelant composition or dilution by the formation water. However, the pre–cross-linked gelling systems may undergo high mechanical retention and filtration through the pores of the reservoir, increasing the pressure losses of the system, and may face operational issues associated with poor injectivity of the swollen particles and risk of plugging nondesired regions of the reservoir matrix.

Several cases of success of field applications of pre–cross-linked polymer systems have been reported for the conformance control of high-permeability zones and anomalies located near-wellbore or deeply into the matrix reservoir rock. The injection of preformed gel particles into the reservoir prevents these systems from substantially invading and damaging the matrix rock adjacent to the treated zone [4, 60].

The major pre–cross-linked polymer systems commercially available or under academic development are presented below. The main differences between them are related to particle sizes, swelling rates, and reservoir conditions in which they can be applied.

2.2.1. Colloidal dispersion gels (CDG)

Colloidal dispersions gels (CDGs) are microgel aggregates that are formed cross-linking a low concentration solution of polymer (e.g. PHPA and PAM-AMPS) with metal ions (e.g. aluminum citrate or chromium citrate). The low polymer concentrations used in these systems are not enough to form a continuous three-dimensional network, and thus, they produce a dispersion of separate gel bundles in which predominate intramolecular cross-links [61–63].

CDG treatments involve the injection of large volumes of microgel aggregates into the formation. Nevertheless, the low polymer concentrations used in their formulation make them cost-effective.

The mechanism by which CDG treatment generates incremental oil production is not fully understood. Laboratory and field tests carried out in the USA and China showed good results for the in-depth conformance control of high-permeability zones of sandstone reservoirs [62].

2.2.2. *Microgel systems*

Microgel systems are micrometer-sized particle gels that are prepared cross-linking a acrylamide terpolymer (with 2% of acrylate and 2% of sulfonate groups) with zirconium (IV) ions and lactate (used as a chelater) or with chromium (III) ion and acetate, under defined shear conditions, to control the microgel globule size (1–10 μm) [4, 64–67].

In these systems, the reservoir permeability control is accomplished by the adsorption of the flexible microgel globules onto the pore walls of the reservoir rock, forming monolayers with thickness equal to the diameter of the gel microspheres. The plugging of the porous medium can be controlled by the appropriate selection of the dimension of the microspheres used for the reservoir treatment.

Some successful field tests were conducted with these flexible microgels for the permeability profile control of Daqing reservoir (China) [68].

2.2.3. *BrightWater™ systems*

BrightWater™ systems are commercial submicro-sized particle gels based on sulfonate-acryl amide-polymers (PAM-AMPS) with both labile and stable cross-links. These systems are prepared using an inverse emulsion polymerization process in order to control the particles size distribution in a sub-micron range (<1 μm). Surfactants are also added into the formulation to fully disperse the micro- or nano-sized particles avoiding them to agglomerate.

These pre–cross-linked systems, developed by MobPTeCh and Ondeo Nalco Energy Services, can be applied for in-depth conformance treatment of reservoirs with temperatures between 35 and 140°C and salinity up to 120,000 mg/L TDS [69, 70].

When BrightWater™ particles are injected into the reservoir, they encounter high temperatures and swell irreversibly (pop up) through the cleavage of the labile cross-links. The cross-linking density of the particle gels decreases allowing them to expand aggressively by absorbing additional water, blocking the high-permeability zones. After the labile cross-links dissociation, the stable cross-links maintain the particle gel network and prevent hydrolysis [69, 70].

After thermal activation, the particles of the BrightWater™ system expand approximately 10 times their original size. The cleavage rate control of the labile cross-links at different temperatures can be made through the appropriate selection of the cross-linking agent used for the submicrogel synthesis [71].

Field tests, both onshore and offshore, carried out in North America, Asia, Africa, Europe, and South America, showed that BrightWater™ submicrogels can travel long distances, allowing the permeability control of the reservoir at great depths [69, 71–73].

2.2.4. *Preformed particle gels (PPG)*

Preformed particle gels (PPG) and partially preformed gel systems are millimeter-sized preformed gels (10 μm to mm) based on acrylamide, polyacrylamide copolymers (e.g. PAM, PHPA, PAtBA, and PAM-AMPS), modified superabsorbent polymers (SAP), or biopolymers (e.g. chitosan and starch) cross-linked with metal ions (e.g. chromium acetate), organic compounds (e.g. N-N′-methylene-bis-acrylamide and PEI) or biopolymers (e.g. chitosan and starch) [74, 75].

The PPGs are prepared by solution polymerization/cross-linking followed by drying, crushing, and sieving the preformed bulk gels to the desired particle size. In the field, prior to be injected into the reservoir, these preformed particles swell in water forming strong and stable hydrogels in surface facilities [74–76].

PPGs were applied in China for more than 2000 wells to reduce the water production in mature water-flooded reservoirs. Since PPGs particles are relatively large, their application is limited to reservoirs with fractures or fracture-like channels.

The PPGs have many advantages over other conformance-improvement treatments, such as [74, 76]:

- good size distribution control of particles (μm–cm);
- insensitivity to reservoir minerals and formation water salinity;
- high thermal and chemical stability at temperatures up to 120°C and salinities up to 300,000 mg/L TDS (including multivalent cations); and
- ease preparation in surface facilities (e.g. produced water can be used to prepare them).

In order to improve the PPG swellability, thermal stability and mechanical strength (modulus, elasticity, and deformability) nanoclays (e.g. kaolinite, laponite, and montmorillonite) can be added to their formulation [75].

2.2.5. *pH-sensitive gel systems*

pH-sensitive gel systems are micrometer-sized particle gels (1–10 μm) that are based on anionic polyelectrolytes (e.g. PHPA and poly(acrylic acid)) cross-linked by allyl ethers of polyols (e.g. allyl pentaerythritol).

In these systems, the pre–cross-linked microgels injection is carried out at low pH values. The pre-addition of hydrochloric acid or citric acid is necessary to reduce the viscosity of the microgels before injection – the low-pH coils polymer chains reducing the gelant viscosity. Once deep into the reservoir, the gel microglobules swell (polymer chains uncoil absorbing formation water) due to the increase of the pH caused by geochemical reactions between the acid (injected with the microgel) and the mineral components of the matrix rock [77].

The microgel swelling-deswelling process occurs due to electrostatic effects caused by adjacent anionic groups that change the hydrodynamic volume of the macromolecules and their conformation in solution [77, 78].

Laboratory and field tests with pH-sensitive gel systems were successfully carried out for in-depth conformance control of reservoirs [60, 79].

3. Gelling polymer system screening

The success of a hydrogel-conformance-improvement treatment depends on the correct assessment of the nature of the conformance problem and on the selection of an effective gelling system.

Reservoir conformance problems that can be treated with gelling polymer systems are basically [21, 22]:

- **Matrix conformance problems**–high-permeability flow path (with no cross-flow) occurring through an unfractured matrix rock with permeability lower than 2 Da. For those, the gelling system is injected into the reservoir, preferably in the gelant state, from the injection well side, improving the overall flood sweep efficiency and generating incremental oil production;
- **Anomaly conformance problems**–high-permeability flow path and/or water/gas coning occurring through anomalies such as: fracture networks (both natural and hydraulically induced), faults, interconnected vugular porosity, caverns, and localized matrix reservoir rock with permeabilities greater than 2 Da. For those, the gelling system is injected into the reservoir, preferably in a matured or partially matured state, from: (i) the production well side for water/gas-shutoff treatments near-wellbore purely blocking the fluid flow, or (ii) the injection well side for the placement of the hydrogels in some significant distance into the fracture or other anomaly surrounding the injection well, functioning as a plugging and diverting agent.

The screening and selection of the most appropriate gelling polymer system for the conformance-improvement treatment of an oil reservoir can be done using different laboratory tests (i.e. bottle tests, continuous and oscillatory rheological measures, and core flooding experimental tests) to access information on the gelation time and final gel strength, as well as the short- and long-term stability of gelling polymer systems under specific reservoir conditions – temperature, salinity, pH value of the formation water, and the presence of either carbon dioxide (CO_2) or hydrogen sulfide (H_2S) [20, 35, 53, 80–83].

Gelling polymer systems used to control the anisotropic permeability profile of oil reservoir should:

- behave as moderately pseudoplastic fluids, with viscosity between 10 and 30 mPa.s at a constant shear rate of 7 s^{-1} to ensure good injectivity and propagation in the porous medium.

These injectivity and propagation parameters can be obtained by continuous rheological tests [56, 84, 85];

- have controllable gelation times, preferably greater than 2 hours, to avoid overloads in the unit's pumping system during the injection of the gelant in the reservoir. Furthermore, the gelation time must be long enough for the gelling system to reach the proposed targets (high-permeability zones or anomalies). This injectivity and propagation parameter parameter can be obtained by bottle tests and continuous and oscillatory rheological tests [45];
- form strong hydrogel under reservoir conditions-with Sydansk's gel-strength code > G (bottle testing, **Figure 3**), elastic modulus (G'), and viscous modulus (G") ratio above 10 (G'/G" > 10) with G' and G" being independent of the oscillation frequency, and G' > 1 Pa. These blocking-ability parameters can be obtained by bottle tests and oscillatory rheological tests [9, 20, 42, 56, 86];
- provide permeability reduction factor (PRF) above 1 after the matrix treatment with the hydrogel. This blocking-ability parameter can be obtained by core flooding experimental tests [78].

Other parameters that should also be considered during the screening are the toxicity and cost of the gelling system components, as well as their thermal, mechanical, and biological stability, retention, and adsorption on reservoir rock.

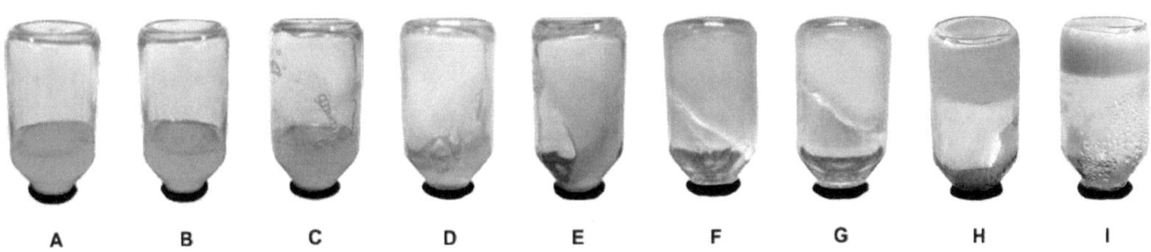

Figure 3. Bottle testing characterization of PAM-PEI hydrogels using the Sydansk's gel-strength code.

4. Conclusion

A large amount of gelling polymer systems applicable for the conformance-improvement treatment of heterogeneous oil reservoirs is commercially available or under academic development.

Choosing the most suitable hydrogel for a conformance problem should be done taking into account the temperature, salinity, and hardness of the reservoir, as well as the pH of the injection water.

Other parameters that should also be considered are the presence of carbon dioxide (CO_2) or hydrogen sulfide (H_2S), the adsorption of the chemicals on reservoir rock, the permeability of the target region, reservoir mineralogy and lithology, as well as toxicity and cost of the gelling system components.

Author details

Fernanda G. C. Tessarolli*, Ailton S. Gomes and Claudia R. E. Mansur

*Address all correspondence to: fernandagcordeiro@gmail.com

Institute of Macromolecules (IMA), Federal University of Rio de Janeiro (UFRJ), Rio de Janeiro, RJ, Brazil

References

[1] Sydansk RD, Romero-Zerón L. Reservoir conformance improvement. Richardson: Society of Petroleum Engineers; 2011

[2] Liu Y, Bai B, Wang Y. Applied technologies and prospects of conformance control treatments in China. Oil and Gas Science and Technology D'IFP Energies Nouvelles. 2010;**65**(6):859-878

[3] Kabir AH. Chemical water & gas shutoff technology: An overview. In: SPE Asia Pacific Improved Oil Recovery Conference, SPE-72119-MS. 2001

[4] Feng Y, Tabary R, Renard M, Le Bon C, Omari A, Chauveteau G. Characteristics of microgels designed for water shutoff and profile control. In: International Symposium on Oilfield Chemistry, SPE-80203-MS. 2003

[5] Hu S, Zhang LH, Yu HJ, Wei W, Luo J. Development and prospect of the profile control/water shutoff technology in reservoir high-capacity channels. Drilling and Production Technology. 2006;**29**(6):117

[6] Ma S, Dong M, Li Z, Shirif E. Evaluation of the effectiveness of chemical flooding using heterogeneous sandpack flood test. Journal of Petroleum Science and Engineering. 2007;**55**(3):294-300

[7] Moradi-Araghi A. A review of thermally stable gels for fluid diversion in petroleum production. Journal of Petroleum Science and Engineering. 2000;**26**(1):10

[8] Sengupta B, Sharma VP, Udayabhanu G. Gelation studies of an organically cross-linked polyacrylamide water shut-off gel system at different temperatures and pH. Journal of Petroleum Science and Engineering. 2012;**81**:145-150

[9] Tokita M, Nishinari K. Gels: structures, properties, and functions: Fundamentals and applications. Vol. 136. Springer-Verlag Belin Heidelberg; 2009

[10] Doe PH, Moradi-Araghi A, Shaw JE, Stahl GA. Development and evaluation of EOR polymers suitable for hostile environments Part 1: Copolymers of vinylpyrrolidone and acrylamide. SPE Reserv Eng SPE-14233-PA. 1987;**2**(04):461-467

[11] Vossoughi S. Profile modification using in situ gelation technology—A review. Journal of Petroleum Science and Engineering. 2000;**26**(1):199-209

[12] Al-Muntasheri GA. Conformance control with polymer gels: What it takes to be successful. Arabian Journal for Science and Engineering. 2012;**37**(4):1131-1141

[13] Gehrke SH. Synthesis, equilibrium swelling, kinetics, permeability and applications of environmentally responsive gels. In: Responsive Gels: Volume Transitions II. Springer; 1993. p. 81-144

[14] Zhao H, Zhao P, Xiao L, Liu X, Bai B. Using associated polymer gels to control conformance for high temperature and high salinity reservoirs. Journal of Canadian Petroleum Technology. 2006;**45**(5):49-54

[15] Vasquez J, Eoff L. Laboratory development and successful field application of a conformance polymer system for low-, medium-, and high-temperature applications. In: SPE Latin American and Caribbean Petroleum Engineering Conference, SPE-139308-MS. Society of Petroleum Engineers; 2010

[16] Seright R, Zhang G, Akanni O, Wang D. A comparison of polymer flooding with in-depth profile modification. Journal of Canadian Petroleum Technology. 2012;**51**(05):393-402

[17] Albonico P, Lockhart TP. Stabilization of polymer gels against divalent ion-induced syneresis. Journal of Petroleum Science and Engineering. 1997;**18**(1):61-71

[18] Donaldson EC, Chilingarian GV, Yen TF. Enhanced oil recovery, I: Fundamentals and analyses. Amsterdam, Netherlands: Elsevier; 1985. e-book ISBN: 9780080868721

[19] Vasquez J. Laboratory evaluation of high-temperature conformance polymer systems. M. Sc. Thesis. University of Oklahoma; 2004

[20] Sydansk RD. A new conformance-improvement-treatment chromium (III) gel technology. In: SPE Enhanced Oil Recovery Symposium. Society of Petroleum Engineers; 1988

[21] Sydansk RD, Southwell GP. More than 12 years' experience with a successful conformance-control polymer-gel technology. SPE Prod Facil SPE-66558-PA. 2000;**15**(04):270-278

[22] Seright RS, Lane RH, Sydansk RD. A strategy for attacking excess water production. In: SPE Permian Basin Oil and Gas Recovery Conference, SPE-70067-MS. Society of Petroleum Engineers; 2001

[23] Seright RS, Liang J. A survey of field applications of gel treatments for water shutoff. In: SPE Latin America/Caribbean Petroleum Engineering Conference, SPE-26991-MS. Society of Petroleum Engineers; 1994

[24] Tackett JE. Hydrolyzed polyacrylamide/chromium (III) acetate gel chemistry. Applied Spectroscopy. 1991;**45**(10):1674-1678

[25] Ricks Jr GV, Portwood JT. Injection-side application of MARCIT polymer gel improves waterflood sweep efficiency, decreases water-oil ratio, and enhances oil recovery in the McElroy Field, Upton County, Texas. In: SPE Permian Basin Oil and Gas Recovery Conference, SPE-59528-MS. Society of Petroleum Engineers; 2000

[26] Portwood JT. The Kansas Arbuckle formation: performance evaluation and lessons learned from more than 200 polymer-gel water-shutoff treatments. In: SPE Production Operations Symposium, SPE-94096-MS. Society of Petroleum Engineers; 2005

[27] Norman C, Turner B, Romero J, Centeno G, Muruaga E. A review of over 100 polymer gel injection well conformance treatments in Argentina and Venezuela: Design, field implementation and evaluation. In: First International Oil Conference and Exhibition in Mexico, SPE-101781-MS. Society of Petroleum Engineers; 2006

[28] Willhite GP, Pancake RE. Controlling water production using gelled polymer systems. SPE Reservoir Evaluation and Engineering SPE-89464-PA. 2008;**11**(03):454-465

[29] Routson WG, Caldwell AL. Method and composition for controlling flow through subterranean formations. US Patent 3,701,384; 1972

[30] Dovan HT, Hutchins RD. Development of a new aluminum/polymer gel system for permeability adjustment. SPE Reservoir Evaluation and Engineering SPE-12641-PA. 1987;**2**(02):177-183

[31] Needham RB, Threlkeld CB, Gall JW. Control of water mobility using polymers and multivalent cations. In: SPE Improved Oil Recovery Symposium, SPE-4747-MS. Society of Petroleum Engineers; 1974

[32] Ghazali HA, Willhite GP. Permeability modification using aluminum citrate/polymer treatments: Mechanisms of permeability reduction in sandpacks. In: SPE Oilfield and Geothermal Chemistry Symposium, SPE-13583-MS. Society of Petroleum Engineers; 1985

[33] Lockhart TP, Albonico P. A new gelation technology for in-depth placement of Cr+3/ polymer gels in high-temperature reservoirs. SPE Prod Facil SPE-24194-PA. 1994;**9**(04): 273-279

[34] Albonico P, Burrafato G, Di Lullo A, Lockhart TP. Effective gelation-delaying additives for Cr+ 3/polymer gels. In: SPE International Symposium on Oilfield Chemistry, SPE-25221 MS. Society of Petroleum Engineers; 1993

[35] Stavland A, Nilsson S. Delayed gelation in corefloods using Cr (III)-malonate as a cross-linker. Journal of Petroleum Science and Engineering. 1995;**13**(3):247-258

[36] Cordova M, Cheng M, Trejo J, Johnson SJ, Willhite GP, Liang J-T, et al. Delayed HPAM gelation via transient sequestration of chromium in polyelectrolyte complex nanoparticles. Macromolecules. 2008;**41**(12):4398-4404

[37] Johnson S, Trejo J, Veisi M, Willhite GP, Liang J-T, Berkland C. Effects of divalent cations, seawater, and formation brine on positively charged polyethylenimine/dextran sulfate/ chromium (III) polyelectrolyte complexes and partially hydrolyzed polyacrylamide/ chromium (III) gelation. Journal of Applied Polymer Science. 2010;**115**(2):1008-1014

[38] Abdo MK, Chung HS, Phelps CH, Klaric TM. Field experience with floodwater diversion by complexed biopolymers. In: SPE Enhanced Oil Recovery Symposium, SPE-12642-MS. Society of Petroleum Engineers; 1984

[39] Rivenq RC, Donche A, Nolk C. Improved scleroglucan for polymer flooding under harsh reservoir conditions. SPE Reservoir Engineering. 1992;**7**(01):15-20

[40] Zolfaghari R, Katbab AA, Nabavizadeh J, Tabasi RY, Nejad MH. Preparation and characterization of nanocomposite hydrogels based on polyacrylamide for enhanced oil recovery applications. Journal of Applied Polymer Science. 2006;**100**(3):2096-2103

[41] Salimi F, Sefti MV, Jarrahian K, Rafipoor M, Ghorashi SS. Preparation and investigation of the physical and chemical properties of clay-based polyacrylamide/Cr (III) hydrogels as a water shut-off agent in oil reservoirs. Korean Journal of Chemical Engineering. 2014;**31**(6):986-993

[42] Wang W, Liu Y, Gu Y. Application of a novel polymer system in chemical enhanced oil recovery (EOR). Colloid & Polymer Science. 2003;**281**(11):1046-1054

[43] Morgan JC, Smith PL, Stevens DG. Chemical adaptation and deployment strategies for water and gas shut-off gel systems. In: Royal Chemistry Society's Chemistry in the Oil Industry 6th International Symposium. Ambleside, UK. 1997. p. 14-17

[44] Hutchins RD, Dovan HT, Sandiford BB. Field applications of high temperature organic gels for water control. In: SPE/DOE Improved Oil Recovery Symposium. Society of Petroleum Engineers; 1996

[45] Jayakumar S, Lane R. Delayed crosslink polymer flowing gel system for water shut-off in conventional and unconventional oil and gas reservoirs. In: SPE International Symposium and Exhibition on Formation Damage Control, SPE-151699-MS. 2012

[46] Reddy BR, Eoff L, Dalrymple E, Black K, Brown D, Rietjens M. A natural polymer-based cross-linker system for conformance gel systems. SPE Journal. 2003;**8**(2):99-106

[47] Eoff L, Dalrymple E, Everett D, Vasquez J. Worldwide field applications of a polymeric gel system for conformance applications. SPE Production & Operations. 2007;**22**(2):231-235

[48] ElKarsani KSM, Al-Muntasheri GA, Hussein IA. Polymer systems for water shut-off and profile modification: A review over the last decade. SPE J SPE-163100-PA. 2014;**19**(01):135-149

[49] Al-Muntasheri G, Nasr-El-Din H, Zitha P. Gelation kinetics and performance evaluation of an organically crosslinked gel at high temperature and pressure. SPE Journal. 2008;**13**(3):337-345

[50] Albonico P, Bartosek M, Malandrino A, Bryant S, Lockhart TP. Studies on phenol-formaldehyde crosslinked polymer gels in bulk and in porous media. In: International Symposium on Oilfield Chemistry. 1995. p. 403-415

[51] Moradi-Araghi A. Application of low-toxicity crosslinking systems in production of thermally stable gels. In: SPE/DOE Improved Oil Recovery Symposium, SPE-27826-MS. Society of Petroleum Engineers; 1994

[52] Sengupta B, Sharma VP, Udayabhanu G. In-situ gelation studies of an eco-friendly cross-linked polymer system for water shut-off at high temperatures. Energy Sources Part Recovery Utilization, and Environmental Effects. 2014;**36**(13):1445-1467

[53] Yadav US, Mahto V. Rheological investigations of partially hydrolyzed polyacrylamide–hexamine–hydroquinone gel. International Journal; 2013

[54] Hardy M, Botermans W, Hamouda A, Valdal J, Warren J. The first carbonate field application of a new organically crosslinked water shutoff polymer system. In: SPE International Symposium on Oilfield Chemistry. 1999. p. 361-376

[55] Jia H, Pu W-F, Zhao J-Z, Jin F-Y. Research on the gelation performance of low toxic PEI cross-linking PHPAM gel systems as water shutoff agents in low temperature reservoirs. Industrial and Engineering Chemistry Research. 2010;**49**(20):9618-9624

[56] ElKarsani KSM, Al-Muntasheri GA, Sultan AS, Hussein IA. Performance of PAM/PEI gel system for water shut-off in high temperature reservoirs: Laboratory study. Journal of Applied Polymer Science. 2015;**132**(17)

[57] Vasquez J, Dalrymple ED, Eoff L, Reddy BR, Civan F. Development and evaluation of high-temperature conformance polymer systems. In: SPE International Symposium on Oilfield Chemistry. Society of Petroleum Engineers; 2005

[58] Shaw A. Study of permeability modification by in-situ gelation of KUSP1-monoethylphthalate ester system in porous media. M. Sc. Thesis. Lawrence, KS: University of Kansas; 1995

[59] Fichadia A. Survey of gelation systems and a study of the gelation of KUSP1 by hydrolysis of mono-ethyl phthalate. M.C Thesis. University of Kansas; 1995

[60] Choi SK. A study of a pH-sensitive polymer for novel conformance control applications. Univ Tex Austin; 2005

[61] Smith JE. Performance of 18 polymers in aluminum citrate colloidal dispersion gels. In: SPE International Symposium on Oilfield Chemistry, SPE-28989-MS. Society of Petroleum Engineers; 1995

[62] Chang HL, Sui X, Xiao L, Guo Z, Yao Y, Yiao Y, et al. Successful field pilot of in-depth colloidal dispersion gel (CDG) technology in Daqing Oilfield. SPE Reservoir Evaluation and Enginnering SPE-89460-PA. 2006;**9**(06):664-673

[63] Al-Assi AA, Willhite GP, Green DW, McCool CS. Formation and propagation of gel aggregates using partially hydrolyzed polyacrylamide and aluminum citrate. SPE Journal. 2009;**14**(03):450-461

[64] Chauveteau G, Tabary R, Le Bon C, Renard M, Feng Y, Omari A, et al. In-depth permeability control by adsorption of soft size-controlled microgels. In: SPE European Formation Damage Conference, SPE-82228-MS. Society of Petroleum Engineers; 2003

[65] Chauveteau G, Omari A, Tabary R, Renard M, Rose J. Controlling gelation time and microgel size for water shutoff. In: SPE/DOE Improved Oil Recovery Symposium, SPE-59317-MS. Society of Petroleum Engineers; 2000

[66] Zaitoun A, Tabary R, Rousseau D, Pichery TR, Nouyoux S, Mallo P, et al. Using microgels to shut off water in a gas storage well. In: International Symposium on Oilfield Chemistry, SPE-106042-MS. Society of Petroleum Engineers; 2007

[67] Cozic C, Rousseau D, Tabary R. Broadening the application range of water shutoff/conformance-control microgels: An investigation of their chemical robustness. In: SPE Annual Technical Conference and Exhibition. Society of Petroleum Engineers, SPE-115974-MS; 2008

[68] Xia H, Wang D, Wu W, Jiang H. Effect of the viscoelasticity of displacing fluids on the relationship of capillary number and displacement efficiency in weak oil-wet cores. In: Asia Pacific Oil and Gas Conference and Exhibition, SPE-109228-MS. Society of Petroleum Engineers; 2007

[69] Frampton H, Morgan JC, Cheung SK, Munson L, Chang KT, Williams D. Development of a novel waterflood conformance control system. In: SPE/DOE Symposium on Improved Oil Recovery, SPE-89391-MS. Society of Petroleum Engineers; 2004

[70] Salehi M, Thomas CP, Kevwitch R, Garmeh G, Manrique EJ, Izadi M. Performance evaluation of thermally-activated polymers for conformance correction applications. In: SPE Improved Oil Recovery Symposium, SPE-154022-MS. Society of Petroleum Engineers; 2012

[71] Ohms D, McLeod JD, Graff CJ, Frampton H, Morgan JC, Cheung SK, et al. Incremental-oil success from waterflood sweep improvement in Alaska. SPE Production and Operations SPE-121761-PA. 2010;**25**(03):247-254

[72] Mustoni J, Denyer P, Norman C. Deep conformance control by a novel thermally activated particle system to improve sweep efficiency in mature waterfloods of the San Jorge basin. In: SPE Improved Oil Recovery Symposium, SPE-129732-MS. Society of Petroleum Engineers; 2010

[73] Roussennac BD, Toschi C. Brightwater trial in Salema field (Campos basin, Brazil). In: SPE EUROPEC/EAGE Annual Conference and Exhibition, SPE-131299-MS. Society of Petroleum Engineers; 2010

[74] Liu Y, Bai B, Shuler PJ. Application and development of chemical-based conformance control treatments in China oilfields. In: SPE/DOE Symposium on Improved Oil Recovery, SPE-99641-MS. Society of Petroleum Engineers; 2006

[75] Aalaie J, Vasheghani-Farahani E, Rahmatpour A, Semsarzadeh MA. Effect of montmorillonite on gelation and swelling behavior of sulfonated polyacrylamide nanocomposite hydrogels in electrolyte solutions. European Polymer Journal. 2008;**44**:2024-2031

[76] Bai B, Wei M, Liu Y. Field and lab experience with a successful preformed particle gel conformance control technology. In: SPE Production and Operations Symposium, SPE-164511-MS. Society of Petroleum Engineers; 2013

[77] Al-Anazi H, Sharma M. Use of a pH sensitive polymer for conformance control. In: International Symposium and Exhibition on Formation Damage Control. Society of Petroleum Engineers; 2002

[78] Lalehrokh F, Bryant S. Application of pH-triggered polymers for deep conformance control in fractured reservoirs. In: SPE Annual Technical Conference and Exhibition, SPE-124773-MS. Society of Petroleum Engineers; 2009

[79] Tessarolli FGC, Queirós YGC, Mansur CRE. Evaluation of pH-sensitive hydrogels to control the permeability anisotropy of oil reservoirs. Journal of Applied Polymer Science. 2014;**131**(17)

[80] Al-Muntasheri GA, Hussein IA, Nasr-El-Din HA, Amin MB. Viscoelastic properties of a high temperature cross-linked water shut-off polymeric gel. Journal of Petroleum Science and Engineering. 2007;**55**(1):56-66

[81] Rousseau D, Chauveteau G, Renard M, Tabary R, Zaitoun A, Mallo P, et al. Rheology and transport in porous media of new water shutoff/conformance control microgels. In: SPE International Symposium on Oilfield Chemistry, SPE-93254-MS. Society of Petroleum Engineers; 2005

[82] ElKarsani KSM, Al-Muntasheri GA, Sultan AS, Hussein IA. Gelation of a water-shutoff gel at high pressure and high temperature: Rheological investigation. SPE J SPE-173185-PA; 2014

[83] Martin FD. Mechanical degradation of polyacrylamide solutions in core plugs from several carbonate reservoirs. SPE Formation Evaluation. 1986;**1**(02):139-150

[84] Diaz I, Nava T, Deolarte C, Castillo O, Vasquez J, Cancino V, et al. Successfully controlling unwanted gas production in a highly naturally fractured carbonate reservoir. In: SPE Western Venezuela Section South American Oil and Gas Congress, SPE-163085-MS. Society of Petroleum Engineers; 2011

[85] Stalker R, Graham GM, Oliphant D, Smillie M. Potential application of viscosified treatments for improved bullhead scale inhibitor placement in long horizontal wells-a theoretical and laboratory examination. In: SPE International Symposium on Oilfield Scale. Society of Petroleum Engineers; 2004

[86] Vega I, Morris W, Robles J, Peacock H, Marin A. Water shut-off polymer systems: Design and efficiency evaluation based on experimental studies. In: SPE Improved Oil Recovery Symposium, SPE-129940-MS; 2010

Section 2

Biomedical Application of Hydrogel

Chapter 5

Enhancement of Hydrogels' Properties for Biomedical Applications: Latest Achievements

Ángel Serrano-Aroca

Additional information is available at the end of the chapter

http://dx.doi.org/10.5772/intechopen.71671

Abstract

Currently, there are many hydrogels used in many important biomedical fields such as therapeutic delivery, contact lenses, corneal prosthesis, bone cements, wound dressing, 3D tissue scaffolds for tissue engineering, etc., due to their excellent biocompatibility and water sorption properties. Many of these hydrophilic polymers have been already approved by the US Food and Drug Administration (FDA) for various applications. However, many of their potential uses required for many biomedical applications often are hindered by their low mechanical strength, antimicrobial and/or antifouling activity, biological interactions, water sorption and diffusion, porosity, electrical and/or thermal properties, among others. Thus, new advanced hydrogels have been developed as multicomponent systems in the form of composite or nanocomposite materials, which are expected to exhibit superior properties to increase the potential uses of these materials in the biomedical industry. Even though the great advances achieved so far, much research has to be conducted still in order to find new strategies to fabricate novel hydrogels able to overcome many of these problems.

Keywords: hydrogels, polymers, composites, nanocomposites, biomedical, properties

1. Introduction

Hydrogels are used in many fields of the biomedical industry such as therapeutic delivery [1], intraocular lenses, contact lenses and corneal prosthesis in ophthalmology [2], bone cements for orthopedics [3], wound dressing [4], 3D tissue scaffolds in regenerative medicine [5], etc., due to their excellent properties such as biocompatibility, water sorption and suitable mechanical performance, among others [6]. Many of these hydrogels have been approved by the US Food and Drug Administration (FDA) for diverse applications and are currently produced massively. However, many of their potential uses required for many biomedical applications are sometimes hindered by their low mechanical strength, biological interactions, electrical and/or thermal properties, water sorption and diffusion, antimicrobial and/or antifouling activity, porosity etc. Thus, new advanced hydrogels have been developed and are currently under intensive research to solve all these problems by means of multicomponent polymeric systems or by combination with other materials and/or nanomaterials to form composites or nanocomposites with enhanced required properties.

2. Mechanical properties

The enhancement of mechanical properties is one of the most desirable achievements in the field of hydrogel engineering and many researchers are currently working in this complex scientific area. Most hydrogels possesses very low mechanical properties, especially in the swollen state. Thus, hydrogels can be reinforced through many established kinds of methods and techniques: block copolymers, in which hydrophobic and hydrophilic domains alternate [7], increasing crosslinking density [8], by means of binary systems composed of two or more mixed polymers as interpenetrating polymer networks [9], by plasma grafting of a hydrogel onto a hydrophobic substrate [10–12], self-reinforced composite materials composed of fibers embedded in a matrix of the same polymer [13] and with the sol–gel reaction to produce nanosilica reinforcement [14]. However, more recent studies have shown new procedures to improve the mechanical properties of hydrogels with the incorporation of nanomaterials such as graphene (2010 Nobel Prize in Physics) and their derivatives: carbon nanotubes (CNT) [15], graphene oxide (GO) [16, 17], reduced graphene oxide (rGO) [18], etc.

2.1. IPNs

Interpenetrating polymer networks (IPN) have gained greater attention during last decades, mainly due to their biomedical applications as reinforced polymer networks. In 2003, Gong et al. [19] reported their discovery of a new hydrogel architecture that produced extraordinarily materials with enhanced mechanical properties, which they termed a double-network (DN) hydrogel. The DN hydrogel was originally believed to be an interpenetrating polymer network (IPN) of a soft neutral polymer network within a more highly crosslinked network prepared by a two-step sequential free-radical polymerization. The first step consisted of the synthesis of a highly crosslinked network, and the second step involved swelling this first network with a water soluble monomer that was then polymerized within it. The second polymerization step was conducted with or without adding a crosslinking agent. Thus, the use of an IPN, which consists of two separate but interwoven polymer networks, is a chemical procedure that is often used in polymer science to control, enhance, and/or combine functional properties. These advanced multicomponent polymeric systems are composed of crosslinked polymer networks without any covalent bonds between them, where at least one of them is synthesized and/or crosslinked within the immediate presence of the other. It is important to differentiate between the six basic multicomponent polymeric structures: mechanical blends, graft copolymers, block copolymers, AB-crosslinked copolymer, semi-IPNs and full-IPNs (see **Figure 1**) [20].

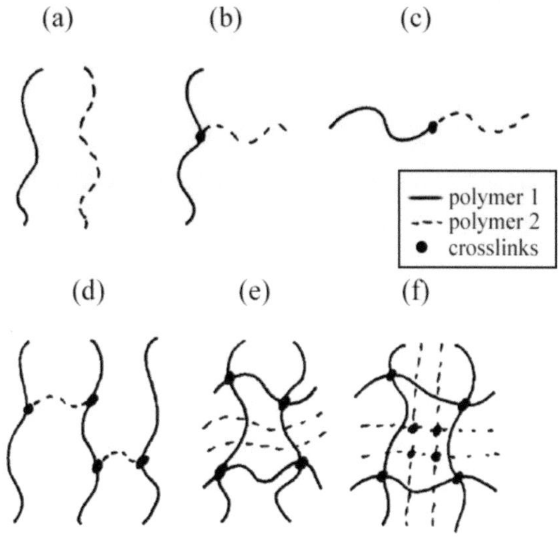

Figure 1. Schematic representation of (a) mechanical blends, (b) graft copolymers, (c) block copolymers, (d) AB-crosslinked copolymer, (e) semi-IPNs and (f) full-IPNs. *Modified from* [20].

If a crosslinker is present in the polymeric system, fully-IPN [21] result, while in the absence of crosslinking, a network having linear polymers embedded within the first crosslinked network is formed (semi- or pseudo-IPN) [22, 23]. IPNs are prepared usually in the form of simultaneous interpenetrating polymer networks (SINs), in which the precursors of both networks are mixed and the two networks are synthesized at the same time, or in the form of sequential IPNs, by swelling of a single-polymer network into a solution containing the mixture of monomer, initiator and activator, usually with a crosslinker. Thus, poly(2-hydroxyethyl methacrylate) (PHEMA) networks were greatly reinforced by poly(ethylene glycol) in SINs and semi-SINs composed of both polymers [24]. Full-IPNs and semi-IPNs of weak gelatin hydrogels were also prepared by the sequential mode of synthesis with polyacrylic acid (PAA) to be evaluated for tissue response in rats [25]. The mechanical properties of pseudo-SIPNs and pseudo-IPNs hydrogels, where the prefix pseudo denotes connectivity of the two network, showed that non-linear tensile properties of pseudo-SIPNs are rate-dependent, but for pseudo-IPNs they are not, which is a consequence of the viscoelastic behavior of a pseudo-SIPN versus elastic performance of the pseudo-IPN [26]. In that study, the mechanical properties of triple-network (TN) hydrogels synthesized from pseudo SIPNs and pseudo-IPNs showed that the presence of a loosely crosslinked third network changes the mechanical behavior of pseudo-SIPNs and pseudo-IPNs by homogenizing the stress within the sample for finite deformations.

IPN hydrogels are also developed with the aim of enhancing the mechanical strength and swelling/deswelling response [27]. For example, interpenetrating polymer network (IPN) hydrogels of chitosan/poly(acrylic acid) (PAA) synthesized by the UV irradiation method showed that even in the swollen state, the present chitosan/PAA IPNs possessed good mechanical properties [28].

"Smart" hydrogels are able to significantly change their volume/shape in response to small alterations of certain parameters of the environment. These responsive hydrogels have numerous applications, being the most of them focused on biological and therapeutic demands [29, 30], and sensing applications [31].

Encapsulation of cells in interpenetrating network (IPN) hydrogels of two biocompatible materials-agarose and poly(ethylene glycol) (PEG) diacrylate with superior mechanical integrity has been developed [32].

Although IPNs based on hydrogels have been extensively reported, the combination of liquid crystalline (LC) property based hydrogels has been rarely explored. In this case, the anisotropic and molecular order of liquid crystals can be combined with the responsive isotropic properties of hydrogels. Thus, advanced stimuli-responsive materials based on interpenetrating liquid crystal-hydrogel polymer networks have been recently fabricated consisting of a cholesteric liquid crystalline network that reflects color and an interwoven poly(acrylic acid) network that provides a humidity and pH response [33].

2.2. Composite hydrogels

Diverse types of chemical modifications of hydrogels do not have a significant change of the overall mechanical strength because of the main structural skeleton of these polymers or copolymers are still weak. However, by fiber reinforcement, the addition of fabrics imparts high strength to the polymer networks, which form the main skeleton of the composites. Thus, fiber-reinforced hydrogels usually consists of a polymer matrix imbedded with high strength fibers, such as glass, aramid and carbon [34]. In such kind of materials, the mechanical properties are presumed to be improved and the biocompatible characteristics of the polymer should remain the same. Thus, hydrogels such as PHEMA, which is one of the most popular biomaterials, has been manufactured by adding various kinds of weaved and knitted fabrics and fibers, in order to improve overall qualities in advanced wound dressing usage [35]. However, in the recent decades, natural fibers as an alternative reinforcement in polymer composites have attracted the attention of many research groups due to their advantages over conventional glass and carbon fibers [36]. These natural fibers include flax, hemp, jute, sisal, kenaf, coir, kapok, banana, henequen and many others, which offer various advantages over man-made glass and carbon fibers such as low cost, low density, comparable specific tensile properties, non-abrasive to the equipment, non-irritation to the skin, reduced energy consumption, less health risk, renewability, recyclability and bio-degradability [37]. Thus, ultra-long chitin natural fibers were incorporated into hydrophobic Poly(methyl methacrylate) (PMMA) to prepare PMMA/Chitin composite hydrogels with improved properties [38]. This achievement is a significantly environmental move toward the sustainable utilization of marine-river crab shell wastes for biomedical applications in good agreement with green chemistry principles.

2.3. Nanocomposite hydrogels

Another alternative and very promising way of reinforcing hydrogels consist of the incorporation of nanomaterials such as silica, graphene and its derivatives, nanofibres or many other nanoparticles. Silica is a biocompatible material which has been reported to possess bioactive properties [39]. Silica can improve the mechanical properties of hydrogels through nanosilica filling or the well-known sol–gel process, which offers a new approach to the synthesis of nanocomposite materials with domain sizes approaching the molecular level [40]. Thus, for example, a biphasic matrix of a hybrid (inorganic–organic) nanocomposite materials of poly(2-hydroxyethyl acrylate) (PHEA) with a silica network obtained by an acid catalyzed sol–gel process of tetraethoxysilane (TEOS) showed a very significant improvement of the mechanical properties of the pure hydrogel [41].

Another reinforcement strategy that can be used consists of the combination of interpenetrated polymer networks and nanosilica filling. Thus, for example, poly(acrylic acid) and alginate IPN material with the incorporation of nanosilica greatly increased the compressive strength of the pure components [42].

Graphene (GN) is a two-dimensional monolayer of sp^2-bonded carbon atoms [43], which has attracted increasing attention in the last decade owing to its excellent electrical and thermal conductivities [44, 45] and great mechanical strength [46]. Furthermore, graphene promotes adherence of human osteoblasts and mesenchymal stromal cells [47], which render this nanomaterial and its derivatives very promising in the biomedical research field. It has been recently reported that its oxidized form, graphene oxide (GO), can greatly enhance the compression performance of alginate hydrogels even in a minuscule concentration [48]. The improvement of mechanical strength of poly(acrylamide) (PAM), which generally exhibit pronounced weakness and brittleness, by incorporating GO to the polymer matrix has also been reported [49]. GO is also a 2D nanomaterial obtained from natural graphite that can be easily exfoliated into monolayer sheets. GO has many hydrophilic oxygenated functional groups, including hydroxyl (–OH), epoxy (–C-O-C-), carbonyl (–C=O) and carboxyl (–COOH), which groups enable its dispersion in water solution [50] and render possible its use in many synthetic procedures. The diversity of unique properties of graphene oxide, including great tensile modulus (1.0 TPa), ultimate strength (130 GPa), electrical and thermal properties [51], render GO an ideal carbon nanomaterial for variety of applications toward the development of new advanced materials. Thus, the addition of GO nanosheets increased the Young's modulus and maximum stress of poly (acrylic acid)/gelatin composite hydrogels significantly as compared with control (0.0 wt. % GO). The highest Young's modulus was observed for hydrogel with GO (0.2 wt. %)/PAA (20 wt. %), whereas the highest maximum stress was detected for GO (0.2 wt. %)/PAA (40 wt. %) specimen. These results suggests that GO nanosheets could be used to improve mechanical properties of hydrogel materials, which are very promising for tissue engineering applications in regenerative medicine [17]. In the biomedical area, a very innovative strategy for three-dimensional self-assembly of graphene oxide sheets and DNA to form multifunctional hydrogels with high mechanical strength, environmental stability, and dye-loading capacity, has also been recently reported [52]. Furthermore, the promising properties of GO with the availability of oxygen-containing functional groups has led the synthesis of 3D crosslinked GO networks able to improve the mechanical properties of alginate hydrogels even more than single GO nanosheets (**Figure 2**) [53].

Other derivatives of graphene, such as carbon nanotubes (CNTs), discovered by Iijima [54], in the form of single wall carbon nanotubes (SWCNTs) and multi-wall carbon nanotubes (MWCNTs), as well as carbon nanofibres (CNFs) are being used for reinforcing and enhancement of many other hydrogels' properties [55–59].

Figure 2. Confocal microscopy of (a) GO/alginate (GO/Alg) (b & d) crosslinked GO/alginate (cGO/Alg), (c) dynamic mechanical analysis (storage modulus (E′) and loss tangent (tan δ)) (e) morphology after swelling in water for 3 hours. *Modified from Ref* [53].

Reinforcement can be also conducted with plant fiber-based nanofibres by a successful fibrillation of wood pulp fibers into nanofiber bundles, which are thin enough to work as well as bacterial cellulose in maintaining the transparency of materials [60]. Other novel nanocomposite hydrogels such as those prepared with polyacrylamide (PAM) as a matrix material reinforced with natural chitosan nanofibres via *in situ* free-radical polymerization showed that these nanofibres acted as a multifunctional cross-linker and a reinforcing agent in the hydrogel polymer system producing a compression strength and a storage modulus significantly higher than those of the pure hydrogel [61].

Other nanoparticles such as clay have been employed to reinforce hydrogels such as polyvinyl alcohol (PVA) [62]. These polymer-clay nanocomposite hydrogels, fabricated for wound healing, constitute a class of materials in which the polymer matrix is reinforced by uniformly dispersed inorganic particles (usually 10 wt.% or less) having at least one dimension in the nanometer scale and exhibiting superior mechanical and thermal properties when compared to pure polymer or conventional composites [63].

3. Electrical properties

The electrical properties are very important in some biomedical areas because it has been demonstrated that various types of electrical stimulation can regulate cell physiological activities such as division [64], migration [65], differentiation and cell death [66]. The electrical stimulation has also been useful in promoting healing for spinal cord repair and cancer therapy due to its non-invasiveness of these polymers [67–69]. Therefore, much emphasis is being done in the development of new conductive hydrogels for biomedical applications. Graphene has been considered to be very effective electrode material due to its excellent conductivity [44], but its production is still very expensive and more new composite materials are expected to be developed with its derivative graphene oxide. However, GO has a very low conductivity due to their oxygen-containing functional groups and must be modified to reduce graphene oxide (rGO) in order to develop electrically conductive hydrogels. Thus, for example, following a single-step procedure starting from a homogeneous water dispersion of GO, it is possible to undergo reduction induced by the UV radiation during the photopolymerization of a resin [70]. Recently, transparent conductive films have been produced by grafting poly(acryl amide)/poly(acrylic acid) on the GO surface followed by a reduction to rGO nanosheets by a two-step chemical reduction with increased conductivity [71] and inorganic–organic double-network (DN) conductive hydrogel of rGO and poly(acrylic acid) has been prepared by a two-step synthesis with a reduction-induced *in situ* self-assembly [72]. Even more recently, a nacre-inspired nanocomposite of rGO and PAA has been prepared *via* a vacuum-assisted filtration self-assembly process (see **Figure 3**). The abundant hydrogen bonding between GO and PAA results in both high strength and toughness of the bioinspired nanocomposites, which are higher than that of pure reduced GO. Moreover, this composite also displays high electrical conductivity, which renders it very promising material in many biomedical applications such as flexible electrodes and artificial muscles.

Carbon nanotubes (CNTs) have also been attracting intensive attention because of their excellent electrical properties with a superb conductivity, remarkable mechanical properties with many potential technological applications [74]. CNTs offer the possibility of developing ultrasensitive electrochemical biosensors due to their unique electrical properties. Thus, nanofibrous membranes filled with multiwalled carbon nanotubes (MWCNT) have been electrospun from the mixture of poly(acrylonitrile-co-acrylic acid) (PANCAA) and MWCNT to develop a glucose biosensor for diabetics [75]. Other hydrogels with very promising biomedical applications consists of dielectrophoretically aligned carbon nanotubes, which control electrical and mechanical properties of gelatin methacrylate (GelMA) hydrogels [76]. The contractile muscle

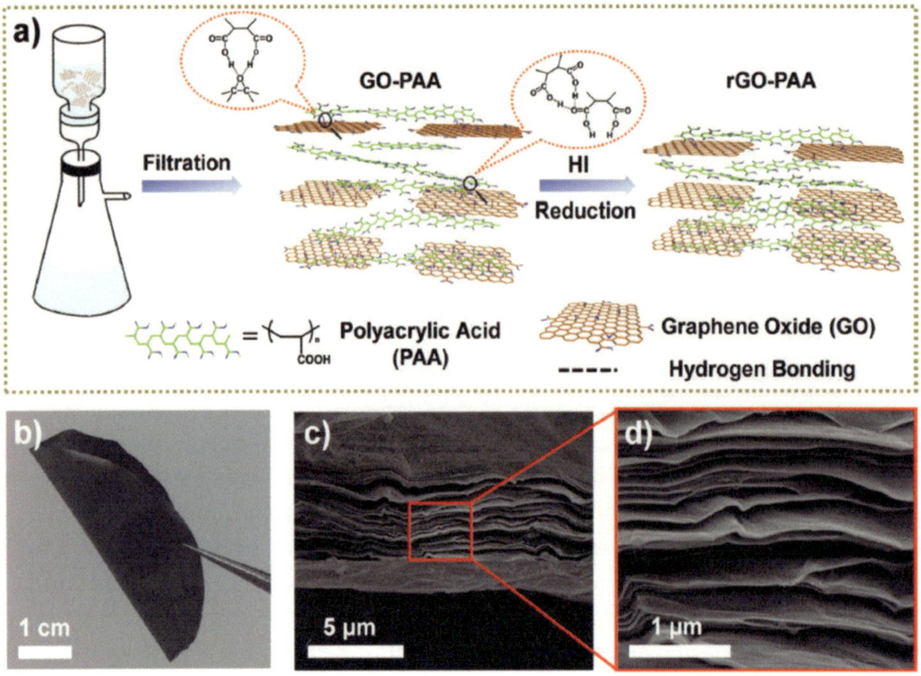

Figure 3. Fabrication process of rGO–PAA composites: (a) the GO nanosheets/PAA homogeneous solution was filtered by vacuum-assisted filtration into GO–PAA composites. Then after hydroiodic acid (HI) reduction, the rGO–PAA composites were obtained. (b) a digital photograph of rGO–PAA composites (c) and (d) cross-section surface morphology with different magnifications of rGO–PAA composites. *Reprinted with permission from Ref.* [73].

cells cultured on these materials demonstrated higher maturation compared with cells cultured on pristine and randomly distributed CNTs in GelMA hydrogels.

4. Thermal properties

Even though hydrogels do not need to endure temperatures higher than that of the human body, the improvement of thermal properties can increase its long-term operation. Thus, for example, the incorporation of polyurethane into polyacrylamide network in the form of an interpenetrating polymer network enhanced the thermal properties due to higher crosslinking density imparted by the hard segment content [22]. Though silica can improve the mechanical properties of hydrogels, the differential scanning calorimetry results of PHEMA/SiO_2 hybrids are complicated, showing two glass transition temperatures (Tg) [77]. However, composite hydrogels with functionalized graphene sheets (FGNS) showed an unprecedented shift in Tg of up to 40 and 30°C in poly(acrylonitrile) with 1 wt. % of this nanomaterial [78].

Another strategy to improve the thermal properties of hydrogels is by nanoparticle filling. Thus, crosslinking metal nanoparticles added into the polymer backbone of PHEMA hydrogels enabled the preparation of thermally stable, soft, magnetic field-driven actuators with muscle-like flexibility [79]. Furthermore, thermal degradation can also be improved by this filling procedure. For example, the mechanical and thermal properties of a renewable and biocompatible hydrogel of gelatin were improved through cross-linking by cellulose nanowhiskers [80].

In the biomedical field, hydrogels are hydrophilic polymers, which are able to absorb large amounts due to contact with cells or tissue in the human body. Therefore, the thermal analysis of water and its influence on the swollen hydrogel properties becomes essential [12, 81, 82].

5. Water sorption and diffusion

Water sorption and diffusion of hydrogels are also very important in biomedical applications because these properties play a very important role in cell survival, especially in tissue engineering [5]. Thus, synthetic hydrogels such as PHEMA or PHEA are very important hydrophilic materials as these polymers were able to absorb and swell retaining large amounts of water within their structure [83–86]. The excellent water sorption property has made these kind of biomaterials very promising in a wide range of biomedical applications such as controlled drug delivery, tissue engineering, wound healing, etc. [6, 87]. The ability of hydrogels to absorb water arises from hydrophilic functional groups attached to the polymeric backbone, while their resistance to dissolution arises from cross-links between network chains [88]. However, these single-network hydrogels have weak mechanical properties in the swollen state and slow response at swelling. Therefore, although reinforcement of hydrogels is absolutely necessary, as already mentioned, the improvement of mechanical properties can significantly affect water sorption. For example, water sorption can be dramatically reduced by the reinforcement produced by the combination of hydrophilic and hydrophobic functional groups of polymers as multicomponent polymeric systems (**Figure 1**).

Reinforcement of hydrogels by GO loading can enhance significantly water sorption and diffusion. Thus, the swelling rates of graphene oxide / poly(acrylic acid-co-acrylamide) nanocomposite hydrogels increased with increase in the GO loadings to 0.30 wt. % and then decreased with further increasing GO loadings. It is worth noting that the hydrogel with only 0.10 wt. % GO exhibited significant improvement of swelling capacity in neutral medium, and could also retain relatively higher swelling rates to a certain degree in acidic and basic solutions. Furthermore, it has been reported very recently that a very low filling of GO can produce a very significant increase of water diffusion (almost 6 times faster) in crosslinked alginate (**Figure 4**) [48]. Therefore, these GO-based super-absorbent hydrogels have very potential applications in many fields such as biomedical engineering and hygienic products [50].

The mechanism of water diffusion [89] can also be altered by the reinforcement of hydrogels through any of the methods shown in Section 2. Thus, very promising biomaterials

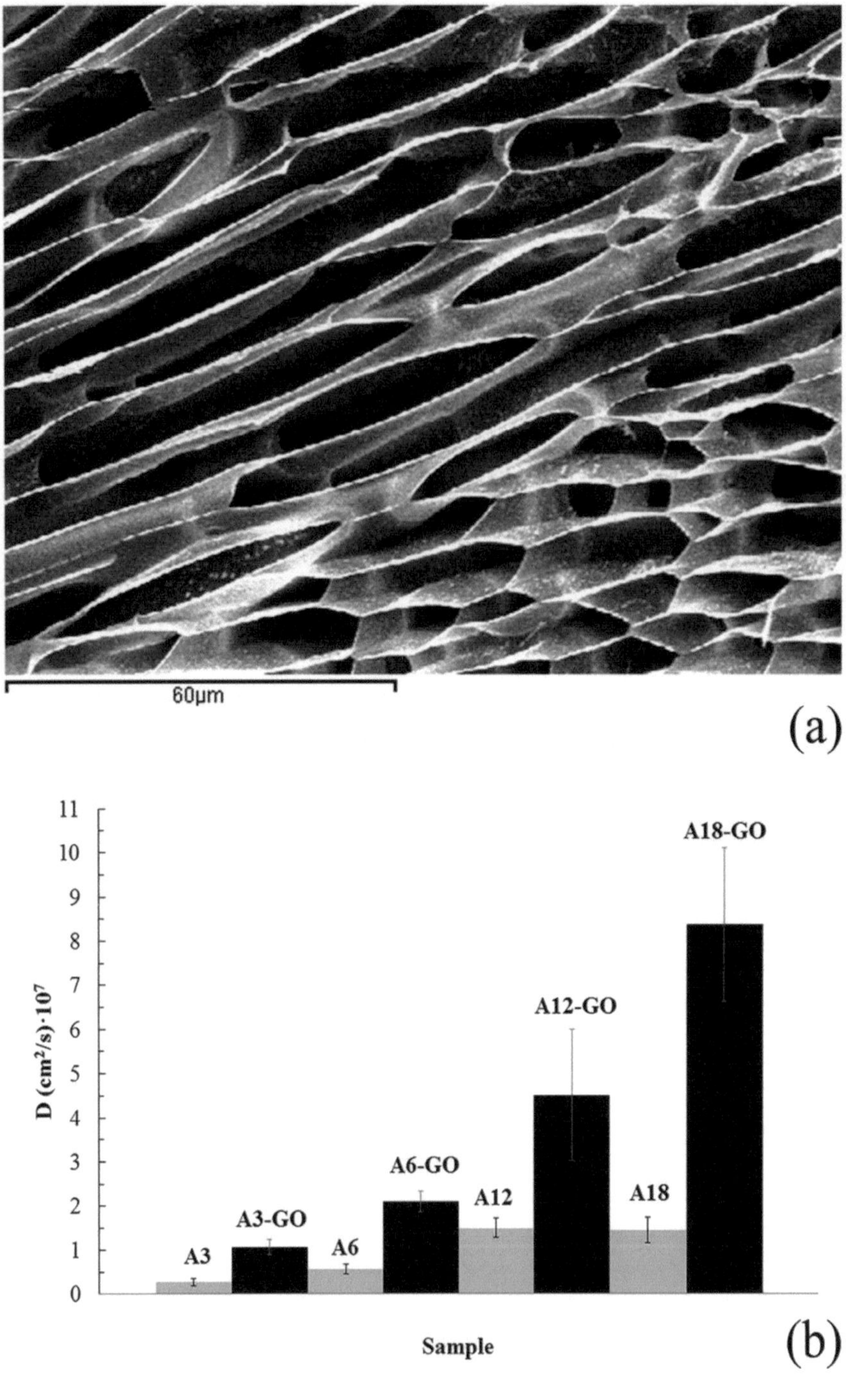

Figure 4. Cryo-scanning electron micrograph of crosslinked alginate synthesized with a minuscule amount of GO and 18 wt.% of calcium chloride (with respect to the mass of sodium alginate) in the swollen state after 2 minutes of immersion in water at 24 ± 0.5°C (a). Apparent diffusion coefficients of liquid water (mean ± standard deviation) in calcium alginate hydrogels with different crosslinker contents with (black columns) and without (gray columns) 0.1 wt.% of GO (b). *Reprinted with permission from Ref* [48].

for drug-releasing such as poly(acrylic acid)-GO composite hydrogels exhibit non-Fickian anomalous diffusion and their deswelling ratio decreases with increasing GO content [51]. Superabsorbent polymers of sodium lignosulfonate-grafted poly(acrylic acid-co-acrylamide), prepared by a new ultrasound synthetic method, shows also a non-Fickian water diffusion transport with a maximum water absorbency of 1350 $g \cdot g^{-1}$ [90]. PHEA hydrogels exhibit also a non-Fickian diffusion behavior s [83, 86]. However, other polymer chemically very similar, PHEMA, which is a very important water-swellable biomedical polymer, is controlled by Fickian diffusion [91]. Thus, copolymerized hydrogels based on 2-hydroxyethyl methacrylate (HEMA) and epoxy methacrylate (EMA) synthesized by bulk polymerizations showed that the swelling process of these polymers is also Fickian and the equilibrium water content (EWC) decreased with increasing EMA content due to its hydrophobicity [92].

It is remarkable that pH has a big influence in the swelling properties and diffusion mechanism of hydrogels. Thus, the swelling properties of semi-interpenetrating polymer networks of acrylamide-based polyurethanes decreased in acidic pH while a reverse trend was observed in basic pH. Nevertheless, these semi-IPNs were found to be hydrolytically stable in phosphate buffer solution, which render them potential materials for biomedical applications [22]. PAA is a pH-sensitive and biocompatible polymer that is being used in many biomedical fields [30] and has attracted considerable interest because of its therapeutic use, due to its ability to swell reversibly with changes in pH. Thus, GO functionalized with PAA (GO-PAA) by *in situ* atom transfer radical polymerization (ATRP) showed potential use as an intracellular protein carrier using bovine serum albumin (BSA) as a model protein [93]. This application is very important because proteins participate in all vital body processes and these perform an essential function inside cells as enzymes, transduction signals, and gene regulation. Another pH-sensitive terpolymer hydrogel, poly(acrylamide-co-2-acrylamido-2-methyl-1-propanesulfonic acid-co-acrylamido glycolic acid), with applications in drug release showed a quasi-Fickian diffusion mechanism with partly chain relaxation controlled diffusion. These hydrogels demonstrated a sharp change in its water absorbency and molecular weight between crosslinks of the network with a change in the swelling media pH [94].

The effect of temperature on swelling properties of hydrogels is also very important [92]. Thus, hydrogels can be modified to exhibit fast temperature sensitivity, and improved oscillating swelling-deswelling properties as for example the thermosensitive poly(N-isopropyl acrylamide-co-acrylic acid) hydrogels [95].

6. Antimicrobial and antifouling activity

Microbial infections can lead to implant failure, which may cause major economic losses and suffering among patients despite the use of preoperative antibiotic prophylaxis and the aseptic processing of materials. Therefore, novel antimicrobial materials are urgently in need for medical uses [96]. For that reason, much effort is being done in the development of advanced hydrogels with inherent antimicrobial properties. Thus, syringe-injectable bioadhesive hydrogels prepared from mixing polydextran aldehyde and branched polyethylenimine, able to kill both Gram-negative and Gram-positive bacteria, while sparing human erythrocytes [97] and injectable conductive self-healed hydrogels based on quaternized chitosan-g-polyaniline (QCSP) and benzaldehyde group functionalized poly(ethylene glycol)-co-poly(glycerol sebacate) (PEGS-FA) with antibacterial, anti-oxidant and electroactive dressing for cutaneous wound healing have been developed [98].

Other hydrogels such as chitosan and its derivatives has been widely used as implant coatings for its intrinsic properties such as non-toxic, osteoconductive, pH responsive, anti-microbial, biocompatible and cell adhesive [99, 100]. Nevertheless, the development of new chitosan derivatives or composites with superior antimicrobial activity is still under research. Thus, for example, a novel hydrogel coating produced by electrophoretic co-deposition of chitosan/alkynyl chitosan showed high antibacterial effect against *Escherichia coli* and *Staphylococcus aureus* [101] by disk diffusion method [102] (see **Figure 5**). Antibacterial polymer coating adhered on the surface of medical implants and devices have attracted great interests in the last decades for its ability to reduce implant-associated infections [103, 104].

Antimicrobial hydrogels formed by crosslinking polyallylamine with aldaric acid derivatives exhibited complete kill within 4 hours against *Pseudomonas aeruginosa, Escherichia coli, Staphylococcus aureus* and *Candida albicans* suspended in culture medium [105] and a facile strategy to fabricate antibacterial ultrathin hydrogel films via a layer-by-layer (LbL) technique and "click" chemistry was reported by Wang et al. [106]. This hydrogels consisted of poly[oligo(ethylene glycol)fumarate]-co-poly[dodecyl bis(2-hydroxyethyl)methylammonium fumarate] (POEGDMAM) containing multi-enes and poly[oligo(ethylene glycol)mercapto-succinate] (POEGMS). These ultrathin films exhibited excellent antibacterial activity against both *Staphylococcus aureus* and *Escherichia coli* due to the presence of ammonium groups with long alkyl chains in the POEGDMAM.

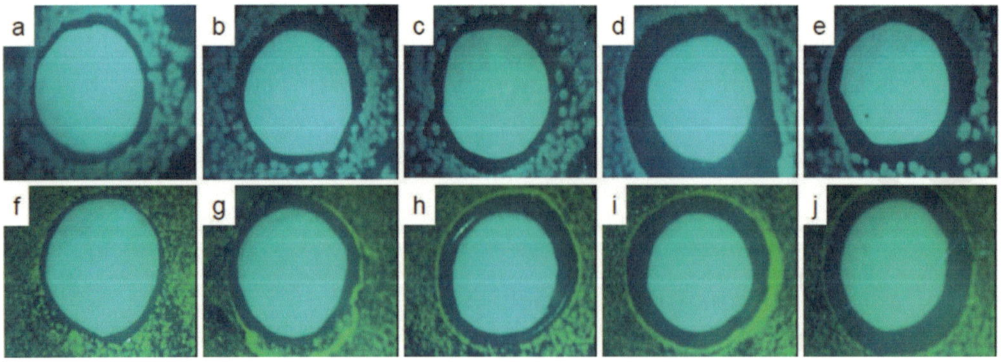

Figure 5. Antimicrobial results of chitosan and alkynyl chitosan against *E. coli* and *S. aureus* by using disk diffusion method. Images (a–e) are paper disks containing 1 wt. % chitosan, 1 wt. % alkynyl chitosan (ACS1, ACS2, ACS3 and ACS4), respectively against *E. coli*; images (f–j) correspond to paper disks containing 1 wt. % chitosan, 1 wt. % alkynyl chitosan (ACS1, ACS2, ACS3 and ACS4), respectively against *S. aureus*. Alkynyl chitosan coded as ACS1, ACS2, ACS3 and ACS4 were prepared by changing the molar ratio of chitosan 0 unit to 3-bromopropyne monomer as 1:0.5, 1:1, 1:1.5 and 1:2. *Reprinted with permission from Ref* [101].

Antibacterial properties can also be imparted to a hydrogel by doping in an exogenous antibiotic for eventual release [107]. In these delivery systems, the active agent is released from the polymer matrix over time. However, the material's antibiotic activity is eventually exhausted with the remaining matrix being left inactive and the remaining vehicle may become a substrate for colonization by bacterial biofilms once the payload is depleted, which can become life threatening. For this reason, secondary surgeries are typically performed to remove these empty depots as a means of preventing this type of infection. To avoid this second surgery, a hydrogel drug delivery system in which the drug release rate of vancomycin and degradation rate of the hydrogel are linked via covalent incorporation of vancomycin in the hydrogel backbone was successfully developed [108].

However, many hydrogels themselves do not have any antimicrobial activity and therefore some fillers, and antimicrobial agents need to be incorporated by physical blending in order to produce antimicrobial materials [109]. Thus, graphene has emerged as a novel green broad-spectrum antimicrobial nanomaterial, with little bacterial resistance and tolerable cytotoxic effect on mammalian cells. It exerts its antibacterial action via physical damages through direct contact of its sharp edges with bacterial membranes and destructive extraction of lipid molecules. The antimicrobial activity of GO against two bacterial pathogens (*Pseudomonas syringae* and *Xanthomonas campestris pv. undulosa*) and two fungal pathogens (*Fusarium graminearum* and *Fusarium oxysporum*) showed that GO had a powerful effect on the reproduction of all these four pathogens because it killed nearly 90% of the bacteria and repressed 80% macroconidia germination along with partial cell swelling and lysis at 500 μgmL^{-1}. The graphene-based nanocomposites have a wide range of biomedical applications, such as wound dressing due to its superior antimicrobial properties and good biocompatibility [110].

Another strategy to design hydrogels with desired antimicrobial performance consists of adding silver nanoparticles (Ag NPs). Hence, silver nanoparticles have emerged up with diverse medical applications ranging from silver based dressings, silver coated medicinal devices, such as nanogels, nanolotions, etc. [111].

Infections are also frequent and highly undesired occurrences after orthopedic procedures. Thus, for example, medicated hydrogels of hyaluronic acid derivatives have been developed [112] to address this problem. However, the growing concern caused by the rise in antibiotic resistance is progressively dwindling the efficacy of such drugs and the integration of silver nanoparticles in hydrogels has become a very promising alternative [113].

The combination of both previous strategies (graphene and Ag NPs) to design antimicrobial hydrogels with good water maintaining ability is of particular significance to promote the development of wound dressing. Thus, a series of hydrogels were synthesized by crosslinking of Ag/graphene composites with acrylic acid and N,N′-methylene bisacrylamide at different mass ratios. In this study, prepared hydrogel with an optimal Ag to graphene mass ratio of 5:1 exhibited much stronger antibacterial abilities than other hydrogels and showed excellent biocompatibility, high swelling ratio, and good extensibility at the same time. Besides, *in vivo* experiments indicated that this nanocomposite hydrogel could significantly accelerate the healing rate of artificial wounds in rats, and it helped to successfully reconstruct intact and thickened epidermis during 15 day of healing of impaired wounds [114]. In the same way, acrylic acid (AA) grafted onto poly(ethylene terephthalate) (PET) film through gamma-ray induced graft copolymerization with silver nanoparticles on the surface showed strong and stable antibacterial activity [115].

It is highly desired to have a hydrogel material bearing the excellent antifouling property/biocompataiblity to prolong the lifetime of implanted materials, switchable antimicrobial property to eliminate infection and inflammation, and good mechanical properties to avoid the failure of the implanted material. We hypothesize derivatives of zwitterionic carboxybetaine with hydroxyl group(s) can switch between the lactone form (anti-microbial) and the zwitterionic form (anti-fouling) and the intramolecular hydrogen bonds will enhance the mechanical property of the zwitterionic hydrogel. It is highly desired to have a hydrogel material bearing the excellent antifouling property/biocompataiblity to prolong the lifetime of implanted materials, switchable antimicrobial property to eliminate infection and inflammation, and good mechanical properties to avoid the failure of the implanted material. We hypothesize derivatives of zwitterionic carboxybetaine with hydroxyl group(s) can switch between the lactone form (anti-microbial) and the zwitterionic form (anti-fouling) and the intramolecular hydrogen bonds will enhance the mechanical property of the zwitterionic hydrogel.

On the other hand, the surface of hydrogels must be modified to make it resistant to protein adsorption and cell adhesion to avoid fouling. Thus, there is a need for coatings with antifouling properties that are able to improve the performances of implanted biomedical devices. Thus, the antifouling properties of poly(2-hydroxyethyl methacrylate-co-methyl methacrylate) hydrogels were improved by the surface grafting of a brush of poly(oligoethylene glycol methyl ether acrylate) [poly(OEGA)] [116]. Novel antifouling highly wettable hydrogels with superior mechanical and self-healing properties have also been developed by UV-initiated copolymerization of non-fouling zwitterionic carboxybetaine methacrylamide (CBMAA-3) and 2-hydroxyethyl methacrylate (HEMA) in the presence of uniformly dispersed clay nanoparticles (Laponite XLG) in water [117].

Therefore, it would be highly desired to have a hydrogel material bearing the excellent antifouling property/biocompataiblity to prolong the lifetime of implanted materials, switchable antimicrobial property to eliminate infection and inflammation, and good mechanical properties to avoid the failure of the implanted material. Thus, derivatives of zwitterionic carboxybetaine have been developed with hydroxyl group(s), which can switch between the lactone form (antimicrobial) and the zwitterionic form (anti-fouling) [118]. Besides, the intramolecular hydrogen bonds enhance the mechanical property of the zwitterionic hydrogel.

Nevertheless, the rapid emergence of antibiotic resistance in pathogenic microbes is becoming an imminent global public health problem because they are highly prone to develop resistance through mutation and the treatment with conventional antibiotics often leads to resistance development leaving the bacterial morphology intact. Therefore, much research is currently being done in the development of new antimicrobial hydrogels because they have been demonstrated to be very effective in preventing and treating multidrug-resistant infections.

7. Porosity

Porous polymers have received an increased level of research interest because of their potential to merge the properties of both porous materials and polymers [119]. Porous polymers have potential applications in many fields such as gas storage and separation materials [120, 121], drug delivery [122], catalysts [123], supports for electrochemical sensing [124], low-dielectric constant materials [125], packing materials in chromatography [126], scaffolds or three-dimensional porous matrices for tissue engineering in regenerative medicine [5, 41, 127, 128] and many others. These high value applications have driven much emphasis on development of reliable methods for preparation of porous polymers with designed pore architectures in the last decades (see **Figure 6**).

Tissue engineering holds great promise for regeneration and repair of diseased tissues, making the development of new porous supports as scaffolds for tissue regeneration a topic of great interest in biomedical research. Hydrogels have emerged as leading candidates for engineered tissue scaffolds due to their good biocompatibility and similarities to native extracellular

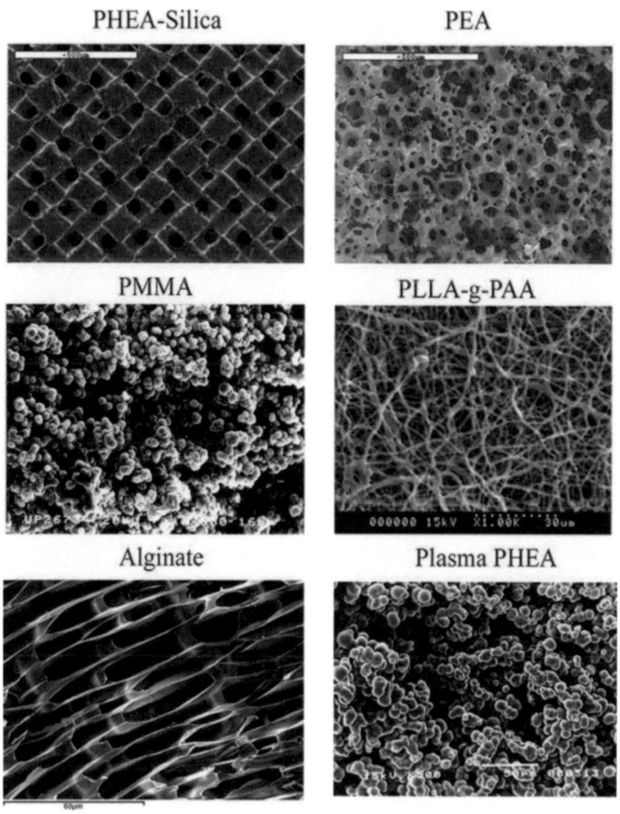

Figure 6. Different porous polymers prepared by different preparation methods to obtain diverse pore architectures. Reprinted with permission from Ref [12, 41, 48, 128–130].

matrix. However, precise control of hydrogel properties, such as high porosity, remains a challenge. Traditional techniques for creating bulk porosity in polymers have demonstrated success in hydrogels for tissue engineering. However, some problems related to direct cell encapsulation often occur. Thus, emerging technologies have demonstrated the ability to control porosity and morphology in hydrogels, creating engineered tissues with structure and function similar to native tissues [131]. The interconnection and geometry of pores, which depends on the tissue to regenerate, physicochemical properties, and mechanical resistance of the material play a major role in these biomedical applications. Thus, there are several methods to produce *scaffolds*, which include gas foaming [132], sintering fiber meshes [133], solvent casting [134], polymerization in solution [86, 135], porogen technique [129, 136], freeze-drying techniques

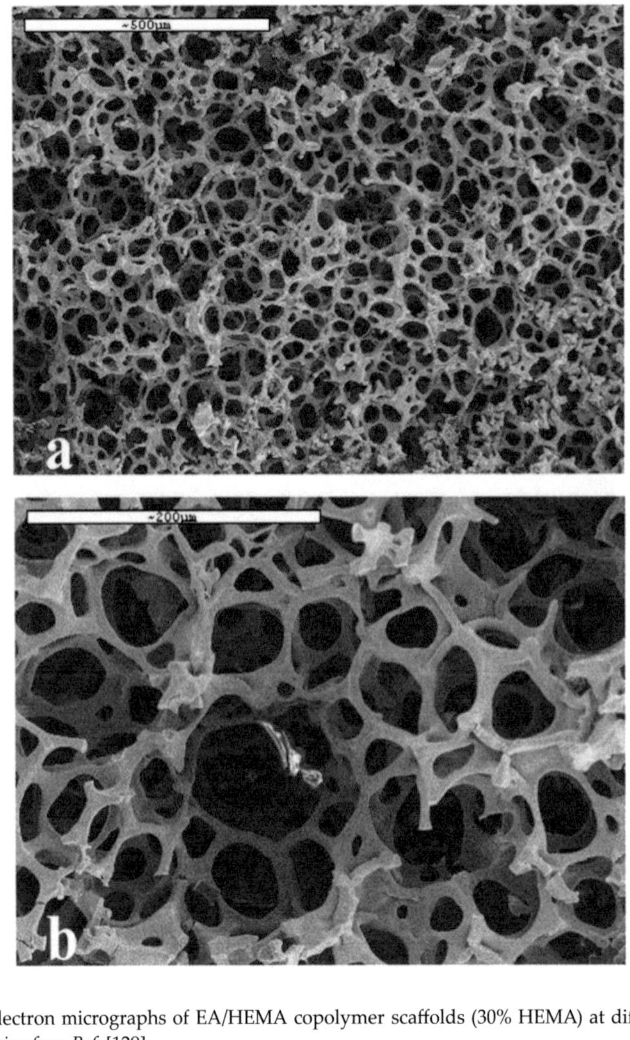

Figure 7. Scanning electron micrographs of EA/HEMA copolymer scaffolds (30% HEMA) at different magnifications. *Reprinted with permission from Ref.* [129].

[137, 138], electrospinning [139], 3D printing [140], 3D bioplotting of scaffold with cells [141], etc. For example, scaffolds with interconnected spherical pores and controlled hydrophilicity with interconnected porous structure were synthesized using a template of sintered PMMA microspheres of controlled size. In these scaffolds, the geometric characteristics (pore size, connectivity and porosity) and the physico-chemical properties of the resulting material can be controlled in an independent way. Copolymerization of hydrophobic ethyl acrylate (EA) and hydrophilic hydroxyethyl methacrylate comonomers in the free space of the template and subsequent solution of the PMMA microspheres gave rise to the scaffold with the designed pore architecture (see **Figure 7**) [129].

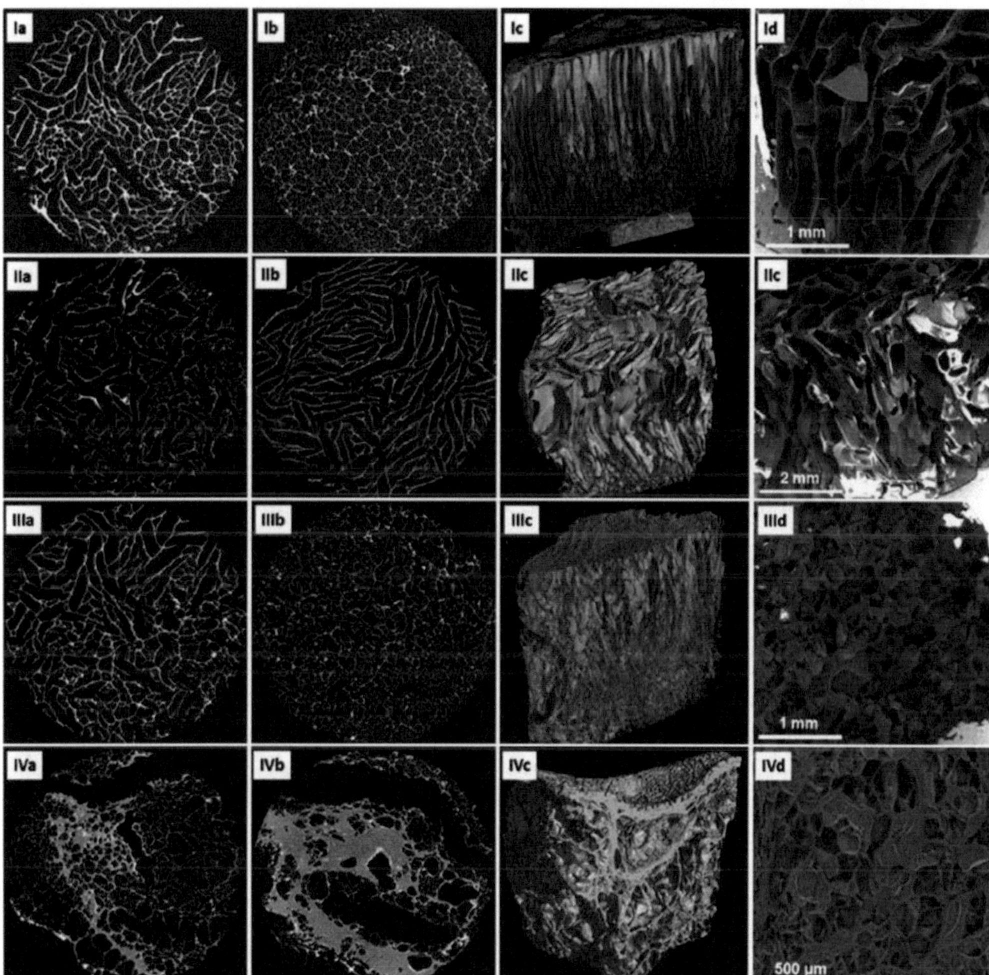

Figure 8. Morphology of the gelatin-PHEMA porous scaffolds as obtained through µ-CT (panels (a)–(c): (a)—top view, (b)—bottom view, and (c)—side view) and SEM (panel (d)) analyses: panel (I)—C0; panel (II)—C1; panel (III)—C2; panel (IV)—C3. *Reprinted with permission from Ref.* [138].

A novel preparation of gelatin-PHEMA porous scaffolds by freeze-drying technique was developed recently [138]. Their morphology was assessed by SEM and μ-CT (**Figure 8**). In this study, four types of novel hydrogels using different methacrylamide-modified gelatin/2-hydroxyethyl methacrylate ratios between 1/0 and 1/2 (w/w) (samples from C0 to C3) were prepared and the results indicated that the HEMA content in the initial polymerization mixtures modulate the architecture of the porous scaffolds from straightforward, top-to-bottom oriented channels for hydrogels possessing the lowest HEMA content to a complex and dense internal porosity of the channels the case of higher HEMA loaded materials. Besides, the covalently bound gelatin sequences significantly improve the biocompatibility of PHEMA based hydrogels, which is very desirable for tissue engineering purposes.

Superporous scaffolds can be also prepared by the salt-leaching technique using NaCl or $(NH4)_2SO_4$ as a porogen [142] or with many other porogenic agents such as ammonium oxalate crystals [143].

By submitting carbon dioxide to supercritical conditions after certain time and then rapidly depressurized is also possible to fabricate porous structures that are related to the supercritical parameters and to the polymer blend composition [131]. The use of CO_2 to create such scaffolds has received some attention in the past but many researchers believe that there is limited interconnectivity between the pores, which is required for tissue engineering. However, highly porous (greater than 85%) and well interconnected scaffolds with very promising applications for cartilage repair have been obtained in a blend of poly(ethyl methacrylate) and tetrahydrofurfuryl methacrylate [144].

Probably the most sophisticated techniques to produce scaffolds are electrospinning, 3D printing and bioprinting. Electrospinning is composed of a high-voltage DC power supply, an infusion pumps and a syringe with a needle tip usually with a diameter of 0.5 mm. For example, a three-dimensional aligned nanofibers-collagen type I hydrogel scaffold for controlled non-viral drug/gene delivery to direct axon regeneration in spinal cord injury treatment has been reported very recently [145].

3D printing promises to produce complex biomedical devices according to computer design using patient-specific anatomical data. This 3D printing technique has slowly evolved to create one-of-a-kind devices, implants, scaffolds for tissue engineering, and drug delivery systems among other important applications. However, several technological limitations, related to the kind of commercially printable materials available and other technical printing aspects such as printing speed, must still be overcome. The common 3D printing technologies are three-dimensional printing, fused deposition modeling, selective laser sintering, stereolithography, and 3D plotting/direct-write/bioprinting, and are still under deep research for the progress of each technology applied in tissue engineering. Bioprinting is the more advanced 3D printing technology because it consists of printing cells combined with custom 3D scaffolds for personalized regenerative medicine [140].

Mechanical resistance depends both on the material properties and on the interconnected pore structure of the scaffold. This problem is more important in polymer hydrogels, which exhibit even poorer mechanical properties when they are porous and hydrated. Therefore, it is usually necessary to enhance the mechanical properties of these porous structures by means of the methods, shown in chapter 2, with nanomaterials or other techniques. For example, the use of a hybrid hydrogel nanocomposite of silica/PHEA as scaffold material matrix greatly improves the mechanical properties [41].

Other modifications of scaffolds such as those of PHEMA with cholesterol methacrylate (CHLMA) and laminin have been developed in the presence of ammonium oxalate crystals to introduce interconnected superpores in the matrix in order to design superporous scaffolds that promote cell-surface interaction [146]. PHEMA has also been modified with laminin-derived Ac-CGGASIKVAVS-OH peptide sequences to construct scaffolds that promote cell adhesion and neural differentiation. With the same goal, nanofiber scaffolds of poly (L-lactide) (PLLA) prepared by electrospinning were treated with oxygen plasma and then simultaneously *in situ* grafted with hydrophilic acrylic acid to obtain PLLA-g-PAA with a modified surface, which significantly improved cell adhesion and proliferation [130].

Polysaccharide hydrogels have become increasingly studied as matrices in soft tissue engineering due to their known cytocompatibility. For example, cross-linkable dextran methacrylates and hyaluronan methacrylate hydrogels, which are candidates as matrices for soft tissue reconstruction, were synthesized showing that the *in vitro* degradation behavior of these types of hydrogels could be controlled by the polysaccharide structure and the cross-linking density. Furthermore, under *in vitro* conditions, these novel materials had no cytotoxic effects against fibroblasts and the use of composite gels improved the adherence of cells [147].

Therefore, great advances have been achieved so far in scaffold design of new advanced porous hydrogels for tissue engineering applications. Nevertheless, much research has to be conducted still in order to find new ways and methods capable of providing suitable materials able to fulfill all the necessary requirements of this biomedical field.

Acknowledgements

This work was supported by the 2017-231-001UCV grant from the Universidad Católica de Valencia "San Vicente Mártir".

Author details

Ángel Serrano-Aroca

Address all correspondence to: angel.serrano@ucv.es

Bioengineering and Cellular Therapy Group, Facultad de Veterinaria y Ciencias Experimentales, Universidad Católica de Valencia San Vicente Mártir, Valencia, Spain

References

[1] Buwalda SJ, Vermonden T, Hennink WE. Hydrogels for therapeutic delivery: Current developments and future directions. Biomacromolecules. 2017;**18**(2):316-330

[2] Chirila T, Harkin D. Biomaterials and Regenerative Medicine in Ophthalmology. 2nd ed. Sawston, Cambridge: Elsevier; 2016

[3] Shalaby S, Nagatomi S, Peniston S. Polymeric Biomaterials for Articulating Joint Repair and Total Joint Replacement. Polymers for Dental and Orthopedic Applications. Boca Raton: CRC Press; 2007

[4] Kamoun EA, Kenawy E-RS, Chen X. A review on polymeric hydrogel membranes for wound dressing applications: PVA-based hydrogel dressings. Journal of Advanced Research. 2017;**8**(3):217-233

[5] Van Blitterswijk C. De Boer J, Tissue Engineering. London: Academic Press; 2014

[6] Ratner BD, Hoffman AS, Schoen FJ, Lemons JE. Biomaterials Science: An Introduction to Materials in Medicine. Canada: Academic Press; 2012

[7] Stoy V, Climent C. In Hydrogels: Speciality Plastics for Biomedical and Pharmaceutical Applications. Basel: Technomic Publishers; 1996

[8] Serrano-Aroca Á, Monleón-Pradas M, Gómez-Ribelles JL. Effect of crosslinking on porous poly(methyl methacrylate) produced by phase separation. Colloid & Polymer Science. 2008;**286**(2):209-216

[9] Ramaraj B, Radhakrishnan G. Modification of the dynamic swelling behaviour of poly(2-hydroxyethyl methacrylate) hydrogels in water through interpenetrating polymer networks (IPNs). Polymer (Guildf). 1994;**35**(10):2167-2173

[10] Serrano-Aroca Á, Monleón-Pradas M, Gómez-Ribelles JL. Plasma-induced polymerisation of hydrophilic coatings onto macroporous hydrophobic scaffolds. Polymer (Guildf). 2007;**48**(7):2071-2078

[11] Serrano-Aroca Á, Gómez-Ribelles JL, Monleón-Pradas M, Vidaurre-Garayo A, Suay-Antón J. Characterisation of macroporous poly(methyl methacrylate) coated with plasma-polymerised poly(2-hydroxyethyl acrylate). European Polymer Journal. 2007;**43**(10):4552-4564

[12] Serrano-Aroca Á, Monleón-Pradas M, Gómez-Ribelles JL, Rault J. Thermal analysis of water in reinforced plasma-polymerised poly(2-hydroxyethyl acrylate) hydrogels. European Polymer Journal. 2015;**72**:523-534

[13] Gilbert JL, Ney DS, Lautenschlager EP. Self-reinforced composite poly (methy1 methacrylate): Static and fatigue properties. 1995;**16**(14):1043-1055

[14] Amerio E, Fabbri P, Malucelli G, Messori M, Sangermano M, Taurino R. Scratch resistance of nano-silica reinforced acrylic coatings. Progress in Organic Coatings. 2008;**62**(2):129-133

[15] Kugler S, Kowalczyk K, Spychaj T. Progress in organic coatings hybrid carbon nanotubes/graphene modified acrylic coats. Progress in Organic Coatings. 2015;**85**:1-7

[16] Cha C, Shin SR, Gao X, Annabi N, Dokmeci MR, Tang XS, et al. Controlling mechanical properties of cell-laden hydrogels by covalent incorporation of graphene oxide. Small. 2014;**10**(3):514-523

[17] Faghihi S, Gheysour M, Karimi A, Salarian R. Fabrication and mechanical characterization of graphene oxide-reinforced poly (acrylic acid)/gelatin composite hydrogels. Journal of Applied Physics. 2014;**115**(83513)

[18] Ha H, Shanmuganathan K, Ellison CJ. Mechanically stable thermally crosslinked poly (acrylic acid)/reduced graphene oxide aerogels. ACS Applied Materials & Interfaces. 2015;**7**(11):6220-6229

[19] Gong JP, Katsuyama Y, Kurokawa T, Osada Y. Double-network hydrogels with extremely high mechanical strength. Advanced Materials. 2003;**15**(14):1155-1158

[20] Sperling LH, Mishra V. The current status of interpenetrating polymer networks. Polymers for Advanced Technologies. 1996;**7**(4):197-208

[21] Maity J, Ray SK. Enhanced adsorption of methyl violet and congo red by using semi and full IPN of polymethacrylic acid and chitosan. Carbohydrate Polymers. 2014;**104**(1):8-16

[22] Merlin DL, Sivasankar B. Synthesis and characterization of semi-interpenetrating polymer networks using biocompatible polyurethane and acrylamide monomer. European Polymer Journal. 2009;**45**(1):165-170

[23] Wang M, Pramoda KP, Goh SH. Mechanical behavior of pseudo-semi-interpenetrating polymer networks based on double-C 60 -end-capped poly (ethylene oxide) and poly (methyl methacrylate). Chemistry of Materials. 2004;**16**:3452-3456

[24] LQ X, Yao F, GD F, Kang ET. Interpenetrating network hydrogels via simultaneous "click chemistry" and atom transfer radical polymerization. Biomacromolecules. 2010;**11**(7):1810-1817

[25] Burugapalli K, Koul V, Dinda AK. Effect of composition of interpenetrating polymer network hydrogels based on poly(acrylic acid) and gelatin on tissue response: A quantitative in vivo study. Journal of Biomedical Materials Research. Part A. 2004;**68**(2):210-218

[26] Shams Es-haghi S, Weiss RA. Finite strain damage-elastoplasticity in double-network hydrogels. Polymer (United Kingdom). 2016;**103**:277-287

[27] Dragan ES. Design and applications of interpenetrating polymer network hydrogels. A review. Chemical Engineering Journal. 2014;**243**:572-590

[28] Lee J, Kim S, Kim S, Lee Y, Lee K. Synthesis and characteristics of interpenetrating polymer network hydrogel composed of chitosan and poly (acrylic acid). Polymer (Guildf). 1998:113-120

[29] Bajpai AK, Shukla SK, Bhanu S, Kankane S. Responsive polymers in controlled drug delivery. Progress in Polymer Science. 2008;**33**(11):1088-1118

[30] Zhao Y, Kang J, Tan T. Salt-, pH- and temperature-responsive semi-interpenetrating polymer network hydrogel based on poly(aspartic acid) and poly(acrylic acid). Polymer (Guildf). 2006;**47**(22):7702-7710

[31] Richter A, Paschew G, Klatt S, Lienig J, Arndt K-F, Adler H-JP. Review on hydrogel-based pH sensors and microsensors. Sensors. 2008;**8**(1):561-581

[32] DeKosky BJ, Dormer NH, Ingavle GC, Roatch CH, Lomakin J, Detamore MS, et al. Hierarchically designed agarose and poly(ethylene glycol) interpenetrating network hydrogels for cartilage tissue engineering. Tissue Engineering. Part C, Methods. 2010;**16**(6):1533-1542

[33] Stumpel JE, Gil ER, Spoelstra AB, Bastiaansen CWM, Broer DJ, Schenning APHJ. Stimuli-responsive materials based on interpenetrating polymer liquid crystal hydrogels. Advanced Functional Materials. 2015;**25**(22):3314-3320

[34] Groover MP. Fundamentals of Modern Manufacturing_Materials, Processes, and Systems. 4th ed. United States of America: J. Wiley & Sons; 2010

[35] Young CD, JR W, Tsou TL. High-strength, ultra-thin and fiber-reinforced pHEMA artificial skin. Biomaterials. 1998;**19**(19):1745-1752

[36] Saheb DN, Jog JP. Natural fiber polymer composites: A review. Advances in Polymer Technology. 1999;**18**(4):351-363

[37] Ku H, Wang H, Pattarachaiyakoop N, Trada M. A review on the tensile properties of natural fiber reinforced polymer composites. Composites. Part B, Engineering. 2011;**42**(4):856-873

[38] Chen C, Li D, Hu Q, Wang R. Properties of polymethyl methacrylate-based nanocomposites: Reinforced with ultra-long chitin nanofiber extracted from crab shells. Materials and Design. 2014 Apr;**56**:1049-1056

[39] Kokubo T. Design of bioactive bone substitutes based on biomineralization process. Materials Science and Engineering: C. 2005;**25**(2):97-104

[40] Hench LL, West JK. The sol-gel process. Chemical Reviews. 1990;**90**(1):33-72

[41] Rodríguez-Hernández JC, Serrano-Aroca Á, Gómez-Ribelles JL, Monleón-Pradas M. Three-dimensional nanocomposite scaffolds with ordered cylindrical orthogonal pores. Journal of Biomedical Materials Research - Part B Applied Biomaterials. 2008;**84**(2):541-549

[42] Lin H-R, Ling M-H, Lin Y-J. High strength and low friction of a PAA-alginate-silica hydrogel as potential material for artificial soft tissues. Journal of Biomaterials Science. Polymer Edition. 2009;**20**(5-6):637-652

[43] Novoselov KS, Geim AK, Morozov SV, Jiang D, Zhang Y, Dubonos SV, et al. Electric field effect in atomically thin carbon films. Science. 2004;**306**(5696):666-669

[44] Geim AK, Novoselov KS. The rise of graphene. Nature Materials. 2007;**6**:183-191

[45] Balandin AA, Ghosh S, Bao W, Calizo I, Teweldebrhan D, Miao F, et al. Superior thermal conductivity of single-layer graphene. Nano Letters. 2008;**8**(3):902-907

[46] Lee C, Wei X, Kysar JW, Hone J. Measurement of the elastic properties and intrinsic strength of monolayer graphene. Science. 2008;**321**(5887):385-388

[47] Kalbacova M, Broz A, Kong J, Kalbac M. Graphene substrates promote adherence of human osteoblasts and mesenchymal stromal cells. Carbon N Y. 2010;**48**(15):4323-4329

[48] Serrano-Aroca Á, Ruiz-Pividal JF, Llorens-Gámez M. Enhancement of water diffusion and compression performance of crosslinked alginate with a minuscule amount of graphene oxide. Scientific Reports. 2017;**7**:11684

[49] Shen J, Yan B, Li T, Long Y, Li N, Ye M. Study on graphene-oxide-based polyacrylamide composite hydrogels. Composites. Part A, Applied Science and Manufacturing. 2012;**43**(9):1476-1481

[50] Huang Y, Zeng M, Ren J, Wang J, Fan L, Xu Q. Preparation and swelling properties of graphene oxide/poly(acrylic acid-co-acrylamide) super-absorbent hydrogel nanocomposites. Colloids Surfaces A Physicochemical and Engineering Aspects. 2012;**401**:97-106

[51] Shen J, Yan B, Li T, Long Y, Li N, Ye M. Mechanical, thermal and swelling properties of poly(acrylic acid)-graphene oxide composite hydrogels. Soft Matter. 2012;**8**(6):1831-1836

[52] Xu Y, Wu Q, Sun Y, Bai H, Shi G. Three-dimensional self-assembly of graphene oxide and DNA into multifunctional hydrogels. ACS Nano. 2010;**4**(12):7358-7362

[53] Serrano-Aroca Á, Deb S. Synthesis of irregular graphene oxide tubes using green chemistry and their potential use as reinforcement materials for biomedical applications. PLoS One. 2017 Sep 21;**12**(9):e0185235

[54] Iijima S. Helical microtubules of graphitic carbon. Nature. 1991;**354**(6348):56-58

[55] Ahadian S, Ramón-Azcón J, Estili M, Liang X, Ostrovidov S, Shiku H, et al. Hybrid hydrogels containing vertically aligned carbon nanotubes with anisotropic electrical conductivity for muscle myofiber fabrication. Scientific Reports. 2014;**4**:4271

[56] Ogoshi T, Takashima Y, Yamaguchi H, Harada A. Chemically-responsive sol-gel transition of supramol single-walled carbon nanotube SWNT hydrogel made by hybrid of SWNTs & cyclodextrin. Journal of the American Chemical Society. 2007;**129**:4878-4879

[57] Satarkar NS, Johnson D, Marrs B, Andrews R, Poh C, Gharaibeh B, et al. Hydrogel-MWCNT nanocomposites: Synthesis, characterization, and heating with radiofrequency fields. Journal of Applied Polymer Science. 2010;**117**(3):1813-1819

[58] Bhattacharyya S, Guillot S, Dabboue H, Tranchant JF, Salvetat JP. Carbon nanotubes as structural nanofibers for hyaluronic acid hydrogel scaffolds. Biomacromolecules. 2008;**9**(2):505-509

[59] Meng X, Stout DA, Sun L, Beingessner RL, Fenniri H, Webster TJ. Novel injectable biomimetic hydrogels with carbon nanofibers and self assembled rosette nanotubes for myocardial applications. Journal of Biomedical Material Research - Part A. 2013;**101 A**(4):1095-1102

[60] Iwamoto S, Nakagaito AN, Yano H, Nogi M. Optically transparent composites reinforced with plant fiber-based nanofibers. Applied Physics A: Materials Science & Processing. 2005;**81**(6):1109-1112

[61] Zhou C, Wu Q. A novel polyacrylamide nanocomposite hydrogel reinforced with natural chitosan nanofibers. Colloids Surfaces B Biointerfaces. 2011;**84**(1):155-162

[62] Kokabi M, Sirousazar M, Hassan ZM. PVA-clay nanocomposite hydrogels for wound dressing. European Polymer Journal. 2007;**43**(3):773-781

[63] Meneghetti P, Qutubuddin S. Synthesis, thermal properties and applications of polymer-clay nanocomposites. Thermochimica Acta. 2006;**442**(1-2):74-77

[64] Zhao M, Forrester JV, McCaig CD. A small, physiological electric field orients cell division. Proceedings of the National Academy of Sciences of the United States of America. 1999;**96**:4942-4946

[65] Yao L, Shanley L, Mccaig C, Zhao M. Small applied electric fields guide migration of hippocampal neurons. Journal of Cellular Physiology. 2008 Aug;**216**(2):527-535

[66] Woo DG, Shim MS, Park JS, Yang HN, Lee DR, Park KH. The effect of electrical stimulation on the differentiation of hESCs adhered onto fibronectin-coated gold nanoparticles. Biomaterials. 2009;**30**(29):5631-5638

[67] Sisken BF, Walker J, Orgel M. Prospects on clinical applications of electrical stimulation for nerve regeneration. Journal of Cellular Biochemistry. 1993;**51**(4):404-409

[68] Zhao M. Electrical fields in wound healing-an overriding signal that directs cell migration. Seminars in Cell & Developmental Biology. 2009;**20**(6):674-682

[69] Kirson ED, Dbalý V, Tovaryš F, Vymazal J, Soustiel JF, Itzhaki A, et al. Alternating electric fields arrest cell proliferation in animal tumor models and human brain tumors. Proceedings of the National Academy of Sciences. 2007;**104**(24):10152-10157

[70] Fabbri P, Valentini L, Bittolo Bon S, Foix D, Pasquali L, Montecchi M, et al. In-situ graphene oxide reduction during UV-photopolymerization of graphene oxide/acrylic resins mixtures. Polymer (Guildf). 2012;**53**(26):6039-6044

[71] Huang YL, Tien HW, Ma CC, Yang SY, SY W, Liu HY, et al. Effect of extended polymer chains on properties of transparent graphene nanosheets conductive film. Journal of Materials Chemistry. 2011;**21**(45):18236

[72] Huang P, Chen W, Yan L. An inorganic-organic double network hydrogel of graphene and polymer. Nanoscale. 2013;**5**(13):6034-6039

[73] Wan S, Hu H, Peng J, Li Y, Fan Y, Jiang L, et al. Nacre-inspired integrated strong and tough reduced graphene oxide-poly(acrylic acid) nanocomposites. Nanoscale. 2016;**8**(10): 5649-5656

[74] Rao CNR, Satishkumar BC, Govindaraj A. Nanotubes. Chemphyschem. 2001;**2**:78-105

[75] Wang ZG, Wang Y, Xu H, Li G, Xu ZK. Carbon nanotube-filled Nanofibrous membranes electrospun from poly(acrylonitrile- co -acrylic acid) for glucose biosensor. Journal of Physical Chemistry C. 2009;**113**(7):2955-2960

[76] Ramón-Azcón J, Ahadian S, Estili M, Liang X, Ostrovidov S, Kaji H, et al. Dielectrophoretically aligned carbon nanotubes to control electrical and mechanical properties of hydrogels to fabricate contractile muscle myofibers. Advanced Materials. 2013;**25**(29): 4028-4034

[77] Huang ZH, Qiu KY. The effects of interactions on the properties of acrylic polymers/ silica hybrid materials prepared by the in situ sol-gel process. Polymer (Guildf). 1997;**38**(3):521-526

[78] Ramanathan T, Abdala AA, Stankovich S, Dikin DA, Herrera-Alonso M, Piner RD, et al. Functionalized graphene sheets for polymer nanocomposites. Nature Nanotechnology. 2008;**3**(6):327-331

[79] Fuhrer R, Athanassiou EK, Luechinger NA, Stark WJ. Crosslinking metal nanoparticles into the polymer backbone of hydrogels enables preparation of soft, magnetic field-driven actuators with muscle-like flexibility. Small. 2009;**5**(3):383-388

[80] Dash R, Foston M, Ragauskas AJ. Improving the mechanical and thermal properties of gelatin hydrogels cross-linked by cellulose nanowhiskers. Carbohydrate Polymers. 2013;**91**(2):638-645

[81] Roorda WE, Bouwstra JA, de Vries MA, Junginger HE. Thermal analysis of water in p(HEMA) hydrogels. Biomaterials 1988;**9**(6):494-499

[82] Megeed Z, Cappello J, Ghandehari H. Thermal analysis of water in silk-elastinlike hydrogels by differential scanning calorimetry. Biomacromolecules. 2004;**5**(3):793-797

[83] Serrano-Aroca Á, Campillo-Fernández AJ, Gómez-Ribelles JL, Monleón-Pradas M, Gallego-Ferrer G, Pissis P. Porous poly(2-hydroxyethyl acrylate) hydrogels prepared by radical polymerisation with methanol as diluent. Polymer (Guildf). 2004;**45**(26):8949-8955

[84] Clayton AB, Chirila TV, Lou X. Hydrophilic sponges based on 2-Hydroxyethyl methacrylate. V. Effect of crosslinking agent reactivity on mechanical properties. Polymer International. 1997;**44**:201-207

[85] Monleón-Pradas M, Gómez-Ribelles JL, Serrano-Aroca Á, Gallego-Ferrer G, Suay-Antón J, Pissis P. Interaction between water and polymer chains in poly(hydroxyethyl acrylate) hydrogels. Colloid & Polymer Science. 2001;**279**(4):323-330

[86] Monleón-Pradas M, Gómez-Ribelles JL, Serrano-Aroca Á, Gallego Ferrer G, Suay Antón J, Pissis P. Porous poly (2-hydroxyethyl acrylate) hydrogels. Polymer (Guildf). 2001;**42**(10):4667-4674

[87] Ahmed EM. Hydrogel: Preparation, characterization, and applications: A review. Journal of Advanced Research. 2015;**6**(2):105-121

[88] Tanaka T. Gels. Scientific American. 1981;**244**(1):124-36, 138

[89] Crank J. The Mathematics of Diffusion Second Edition. New York: Oxford University Press; 1975

[90] Wang X, Zhang Y, Hao C, Dai X, Zhu F, Ge C. Ultrasonic synthesis and properties of a sodium lignosulfonate–grafted poly(acrylic acid-co-acryl amide) composite super absorbent polymer. New Journal of Chemistry. 2014;**38**(12):6057-6063

[91] Gehrke SH, Biren D, Hopkins JJ. Evidence for Fickian water transport in initially glassy poly(2-hydroxyethyl methacrylate). Journal of Biomaterials Science. Polymer Edition. 1995;**6**(4):375-390

[92] Wang J, Wu W. Swelling behaviors, tensile properties and thermodynamic studies of water sorption of 2-hydroxyethyl methacrylate/epoxy methacrylate copolymeric hydrogels. European Polymer Journal. 2005;**41**(5):1143-1151

[93] Kavitha T, Kang IK, Park SY. Poly(acrylic acid)-grafted graphene oxide as an intracellular protein carrier. Langmuir. 2014;**30**(1):402-409

[94] Krishna-Rao KSV, Ha CS. PH sensitive hydrogels based on acryl amides and their swelling and diffusion characteristics with drug delivery behavior. Polymer Bulletin. 2009;**62**(2):167-181

[95] Zhang XZ, Yang YY, Wang FJ, Chung TS. Thermosensitive poly(N-isopropylacrylamide-co-acrylic acid) hydrogels with expanded network structures and improved oscillating swelling-deswelling properties. Langmuir. 2002;**18**(6):2013-2018

[96] Shi L, Chen J, Teng L, Wang L, Zhu G, Liu S, et al. The antibacterial applications of Graphene and its derivatives. Small. 2016;**12**(31):4165-4184

[97] Giano MC, Ibrahim Z, Medina SH, Sarhane KA, Christensen JM, Yamada Y, et al. Injectable bioadhesive hydrogels with innate antibacterial properties. Nature Communications. 2014;**5**(May):1-9

[98] Zhao X, Wu H, Guo B, Dong R, Qiu Y, Ma PX. Antibacterial anti-oxidant electroactive injectable hydrogel as self-healing wound dressing with hemostasis and adhesiveness for cutaneous wound healing. Biomaterials. 2017;**122**:34-47

[99] Jayakumar R, Prabaharan M, Nair SV, Tamura H. Novel chitin and chitosan nanofibers in biomedical applications. Biotechnology Advances. 2010;**28**:142-150

[100] Jayakumar R, Prabaharan M, Sudheesh Kumar PT, Nair SV, Tamura H. Biomaterials based on chitin and chitosan in wound dressing applications. Biotechnology Advances. 2011;**29**:322-337

[101] Ding F, Nie Z, Deng H, Xiao L, Du Y, Shi X. Antibacterial hydrogel coating by electrophoretic co-deposition of chitosan/alkynyl chitosan. Carbohydrate Polymers. 2013;**98**(2):1547-1552

[102] Bauer AW, Kirby WMM, Sherris JC, Turck AM. Antibiotic susceptibility testing by a standardized single disk method. American Journal of Clinical Pathology. 1966;**45**:493-496

[103] Zhou C, Li P, Qi X, Sharif ARM, Poon YF, Cao Y, et al. A photopolymerized antimicrobial hydrogel coating derived from epsilon-poly-l-lysine. Biomaterials. 2011;**32**(11):2704-2712

[104] Li P, Poon YF, Li W, Zhu H-Y, Yeap SH, Cao Y, et al. A polycationic antimicrobial and biocompatible hydrogel with microbe membrane suctioning ability. Nature Materials. 2011;**10**(2):149-156

[105] Andrews MA, Figuly GD, Chapman JS, Hunt TW, Glunt CD, Rivenbark JA, et al. Antimicrobial hydrogels formed by crosslinking polyallylamine with aldaric acid derivatives. Journal of Applied Polymer Science. 2011;**119**(6):3244-3252

[106] Wang H, Zha G, Du H, Gao L, Li X, Shen Z, et al. Facile fabrication of ultrathin antibacterial hydrogel films via layer-by-layer "click" chemistry. Polymer Chemistry. 2014;**5**(22):6489-6494

[107] Smith AW. Biofilms and antibiotic therapy: Is there a role for combating bacterial resistance by the use of novel drug delivery systems? Advanced Drug Delivery Reviews. 2005;**57**:1539-1550

[108] Lakes AL, Peyyala R, Ebersole JL, Puleo DA, Hilt JZ, Dziubla TD. Synthesis and characterization of an antibacterial hydrogel containing covalently bound vancomycin. Biomacromolecules. 2014;**15**(8):3009-3018

[109] He J, Söderling E, Lassila LVJ, Vallittu PK. Incorporation of an antibacterial and radiopaque monomer in to dental resin system. Dental Materials. 2012;**28**(8):e110-e117

[110] Ji H, Sun H, Antibacterial QX. Applications of graphene-based nanomaterials: Recent achievements and challenges. Advanced Drug Delivery Reviews. 2016;**105**:176-189

[111] Rai M, Yadav A, Gade A. Silver nanoparticles as a new generation of antimicrobials. Biotechnology Advances. Elsevier. 2009;**27**:76-83

[112] Pitarresi G, Palumbo FS, Calascibetta F, Fiorica C, Di Stefano M, Giammona G. Medicated hydrogels of hyaluronic acid derivatives for use in orthopedic field. International Journal of Pharmaceutics. 2013;**449**(1-2):84-94

[113] González-Sánchez MI, Perni S, Tommasi G, Morris NG, Hawkins K, López-Cabarcos E, et al. Silver nanoparticle based antibacterial methacrylate hydrogels potential for bone graft applications. Materials Science and Engineering: C. 2015;**50**:332-340

[114] Fan Z, Liu B, Wang J, Zhang S, Lin Q, Gong P, et al. A novel wound dressing based on Ag/graphene polymer hydrogel: Effectively kill bacteria and accelerate wound healing. Advanced Functional Materials. 2014;**24**(25):3933-3943

[115] Ping X, Wang M, Xuewu G. Surface modification of poly(ethylene terephthalate) (PET) film by gamma-ray induced grafting of poly(acrylic acid) and its application in antibacterial hybrid film. Radiation Physics and Chemistry. 2011;**80**(4):567-572

[116] Bozukova D, Pagnoulle C, De Pauw-Gillet MC, Ruth N, Jérôme R, Jérôme C. Imparting antifouling properties of poly(2-hydroxyethyl methacrylate) hydrogels by grafting poly(oligoethylene glycol methyl ether acrylate). Langmuir. 2008;**24**(13):6649-6658

[117] Kostina NY, Sharifi S, de los Santos Pereira A, Michálek J, Grijpma DW, Rodriguez-Emmenegger C. Novel antifouling self-healing poly(carboxybetaine methacrylamide-co-HEMA) nanocomposite hydrogels with superior mechanical properties. Journal of Materials Chemistry B 2013;**1**(41):5644

[118] Cao B, Tang Q, Li L, Humble J, Wu H, Liu L, et al. Switchable antimicrobial and antifouling hydrogels with enhanced mechanical properties. Advanced Healthcare Materials. 2013;**2**(8):1096-1102

[119] Wu D, Xu F, Sun B, Fu R, He H, Matyjaszewski K. Design and preparation of porous polymers. Chemical Reviews. American Chemical Society. 2012;**112**:3959-4015

[120] Du N, Park HB, Robertson GP, Dal-Cin MM, Visser T, Scoles L, et al. Polymer nanosieve membranes for CO2-capture applications. Nature Materials. 2011;**10**(5):372-375

[121] Shetty D, Jahovic I, Raya J, Ravaux F, Jouiad M, Olsen J-C, et al. An ultra-absorbent alkyne-rich porous covalent polycalix[4]arene for water purification. Journal of Materials Chemistry A. 2017;**5**:62

[122] Kolanthai E, Dikeshwar Colon VS, Sindu PA, Chandra VS, Karthikeyan KR, Babu MS, et al. Effect of solvent; enhancing the wettability and engineering the porous structure of a calcium phosphate/agarose composite for drug delivery. RSC Advances. 2015;**5**(24):18301-18311

[123] Pulko I, Wall J, Krajnc P, Cameron NR. Ultra-high surface area functional porous polymers by emulsion templating and hypercrosslinking: Efficient nucleophilic catalyst supports. Chemistry - A European Journal. 2010;**16**(8):2350-2354

[124] Zhao C, Danish E, Cameron NR, Kataky R. Emulsion-templated porous materials (PolyHIPEs) for selective ion and molecular recognition and transport : Applications in electrochemical sensing. Journal of Materials Chemistry. 2007;**17**(23):2446

[125] GD F, Yuan Z, Kang ET, Neoh KG, Lai DM, Huan ACH. Nanoporous ultra-low-dielectric-constant fluoropolymer films via selective UV decomposition of poly(pentafluorostyrene)-block-poly(methyl methacrylate) copolymers prepared using atom transfer radical polymerization. Advanced Functional Materials. 2005;**15**(2): 315-322

[126] Lv Y, Hughes TC, Hao X, Hart NK, Littler SW, Zhang X, et al. A novel route to prepare highly reactive and versatile chromatographic monoliths. Macromolecular Rapid Communications. 2010;**31**(20):1785-1790

[127] Murphy AR, Laslett A, O'Brien CM, Cameron NR. Scaffolds for 3D in vitro culture of neural lineage cells. Acta Biomaterialia. 2017

[128] Serrano-Aroca Á, Llorens-Gámez M. Dynamic mechanical analysis and water vapour sorption of highly porous poly(methyl methacrylate). Polymer (Guildf). 2017;**125**:58-65

[129] Brígido-Diego R, Pérez-Olmedilla M, Serrano-Aroca ÁS, Gómez-Ribelles JL, Monleón-Pradas M, Gallego-Ferrer G, et al. Acrylic scaffolds with interconnected spherical pores and controlled hydrophilicity for tissue engineering. Journal of Materials Science. Materials in Medicine. 2005;**40**(18):4881-4887

[130] Park K, Hyun JJ, Kim JJ, Ahn KD, Dong KH, Young MJ. Acrylic acid-grafted hydrophilic electrospun nanofibrous poly(L-lactic acid) scaffold. Macromolecular Research. 2006;**14**(5):552-558

[131] Annabi N, Nichol JW, Zhong X, Ji C, Koshy S, Khademhosseini A, et al. Controlling the porosity and microarchitecture of hydrogels for tissue engineering. Tissue Engineering. Part B, Reviews. 2010;**16**(4):371-383

[132] Arora KA, Lesser AJ, McCarthy TJ. Compressive behavior of microcellular polystyrene foams processed in supercritical carbon dioxide. Polymer Engineering and Science. 1998;**38**(12):2055-2062

[133] Thomson RC, Wake MC, Yaszemski MJ, Mikos AG. Biodegradable polymer scaffolds to regenerate organs. Advances in Polymer Science. 1995;**122**:247-277

[134] Andrianova GP, Pakhomov SI. Porous materials from crystallizable polyolefins produced by gel technology. Polymer Engineering and Science. 1997;**37**(8):1367-1380

[135] Serrano-Aroca Á, Monleón-Pradas M, Gómez-Ribelles JL. Macroporous poly(methyl methacrylate) produced by phase separation during polymerisation in solution. Colloid & Polymer Science. 2007;**285**(7):753-760

[136] Flynn L, Dalton PD, Shoichet MS. Fiber templating of poly(2-hydroxyethyl methacrylate) for neural tissue engineering. Biomaterials. 2003;**24**(23):4265-4272

[137] Kang HW, Tabata Y, Ikada Y. Fabrication of porous gelatin scaffolds for tissue engineering. Biomaterials. 1999;**20**(14):1339-1344

[138] Dragusin DM, Van Vlierberghe S, Dubruel P, Dierick M, Van Hoorebeke L, Declercq HA, et al. Novel gelatin–PHEMA porous scaffolds for tissue engineering applications. Soft Matter. 2012;**8**(37):9589

[139] Agarwal S, Wendorff JH, Greiner A. Use of electrospinning technique for biomedical applications. Polymer (Guildf). 2008;**49**(26):5603-5621

[140] Chia HN, Recent WBM. Advances in 3D printing of tissue engineering scaffolds. Journal of Biological Engineering. 2015;**9**(4):2-14

[141] Derby B. Printing and prototyping of tissues and scaffolds. Science. 2012;**338**(6109):921-926

[142] Horák D, Hlídková H, Hradil J, Lapčíková M, Šlouf M. Superporous poly(2-hydroxyethyl methacrylate) based scaffolds: Preparation and characterization. Polymer (Guildf). 2008;**49**(8):2046-2054

[143] Kubinová Š, Horák D, Kozubenko N, Vaněček V, Proks V, Price J, et al. The use of superporous ac-CGGASIKVAVS-OH-modified PHEMA scaffolds to promote cell adhesion and the differentiation of human fetal neural precursors. Biomaterials. 2010;**31**(23): 5966-5975

[144] Barry JJA, Silva MMCG, Cartmell SH, Guldberg RE, Scotchford CA, Howdle SM. Porous methacrylate tissue engineering scaffolds: Using carbon dioxide to control porosity and interconnectivity. Journal of Materials Science. 2006;**41**(13):4197-4204

[145] Nguyen LH, Gao M, Lin J, Wu W, Wang J, Chew SY. Three-dimensional aligned nanofibers-hydrogel scaffold for controlled non-viral drug/gene delivery to direct axon regeneration in spinal cord injury treatment. Scientific Reports. 2017;**7**(October 2016):42212

[146] Kubinová Š, Horák D, Syková E. Cholesterol-modified superporous poly(2-hydroxyethyl methacrylate) scaffolds for tissue engineering. Biomaterials. 2009;**30**(27):4601-4609

[147] Möller S, Weisser J, Bischoff S, Schnabelrauch M. Dextran and hyaluronan methacrylate based hydrogels as matrices for soft tissue reconstruction. Biomolecular Engineering. 2007;**24**(5):496-504

Chapter 6

Hydrogels Fibers

Javad Foroughi, Azadeh Mirabedini and
Holly Warren

Additional information is available at the end of the chapter

http://dx.doi.org/10.5772/intechopen.74188

Abstract

With the ever increasing demand for suitable tissue engineering and drug delivery systems, hydrogel fiber spinning has drawn increasing attention due to its ability to create three-dimensional (3D) structures using biomaterials. Hydrogel materials have shown a great promise to be used as templates for tissue engineering and implantable devices. Among the many production techniques available, advanced fiber processing, such as coaxial and triaxial spinning of natural hydrogels, has attracted a great deal of attention because the basic core-sheath structure provides a drug delivery system capable of delivering high concentrations of drug for localized drug delivery and tissue engineering applications. Encapsulating the drug and bioactive cores with a more bio-friendly coating allows for a versatile system for producing devices with appropriate mechanical, chemical and biological properties that can mimic the native extracellular matrix, better supporting cell growth and maintenance. This chapter presents a novel fabrication method using a wet-spinning process that allows for the routine production of multifunctional coaxial hydrogel fibers that take advantage of the encapsulating properties of a hydrogel core while also promoting good cell growth and biocompatibility via the use of bio-friendly material in the sheath.

Keywords: fiber, biomaterial, hydrogels, biomedical applications, drug delivery

1. Introduction

1.1. Material considerations for biomedical applications

For an ideal scaffolding material, properties are required that include biocompatibility, suitable microstructure, desired mechanical strength and degradation rate as well as, most importantly, the ability to support cell residence and allow retention of metabolic functions. Various approaches to engineer tissues currently exist and rely on the implementation of a material scaffold. These structural scaffolds aim to serve as a synthetic extracellular matrix (ECM) to both guide cell growth in a controlled 3D pattern and directly apply certain stimuli which promotes this growth. These scaffolds and stimuli are tailored for specific tissue growth and applications and therefore can vary greatly. A wide range of materials are known to be utilized as cell supporting materials in biomedical applications including natural and synthetic polymers, metals, ceramics and alloys [1]. Aside from the specific materials used in certain applications such as orthopedics, dental implants and artificial vascular materials, the focus of this chapter is on the role of naturally occurring hydrogels to develop biofibres with the final use as biocompatible templates for the purpose of drug delivery systems or the building blocks of tissue scaffolds. Herein, hydrogels including alginate and chitosan are introduced and explained as follows.

1.2. Natural hydrogels

Natural polymers can be considered as the most commonly used clinical biomaterials. Polymers of natural origin are attractive options, mainly due to their similarities with ECM as well as their chemical versatility and biological performance [2, 3]. A variety of hydrogels [4, 5] have been investigated as potential tissue scaffold materials. Hydrogels are three-dimensional, covalently crosslinked polymer networks with a high number of hydrophilic groups, capable of accommodating large amounts of water. Their hydrophilic polymer chains are either synthetic or naturally sourced and can exhibit tunable mechanical properties depending on their cross-linking mechanisms. Hydrogels used in bioapplications are typically degradable, can be processed under relatively mild conditions, have mechanical and structural properties similar to many tissues and the ECM, and can be delivered in a minimally invasive manner.

Hydrogels demonstrated a distinct efficacy as matrices for 3D cell culture since they are mechanically similar to living tissues and ECM, such as a soft and rubbery yet deformable nature, with low interfacial tension with biological fluids [6]. In addition, they can use biologically relevant electrolytes which makes them well suited for applications within biology [7]. Another unique characteristic of biomimetic hydrogels is that they may undergo huge volume changes, which occur in relatively narrow ranges of changes of temperature, pH, and ionic strength [8].

Polysaccharides are a typical group of natural biopolymers showing great swellability that make them ideal candidates for preparing hydrogels. Polysaccharides are high molecular weight polymeric carbohydrates formed from repeating monosaccharide units [1]. Polysaccharides are advantageous for biomedical applications due to their wide availability, low cost as well as the presence of functional groups in the polymer chain. They offer a wide diversity in structure and properties due to their wide range of molecular weight and chemical composition. Alginate and chitosan are considered as the most extensively used gel-forming polysaccharides for cell growth from natural sources. They were chosen and used in this study mainly due to their several unique properties including biodegradability, biocompatibility, low toxicity, promoting attachment, migration, proliferation and differentiation of cells and antimicrobial activity as well as ease of fabrication and availability [1].

1.3. Sodium alginate

Alginate, or alignic acid, is an anionic polysaccharide extracted from brown algae and seaweeds. Discovered by Stanford in 1881, it is a common food additive but also used in

http://dx.doi.org/10.5772/intechopen.74188

pharmaceuticals, textile printing and for many other applications. Alginate is a linear, binary copolymer composed of 1, 4-linked β-D-mannuronic acid (M) and α-L-guluronic acid (G) monomers. Alginates are extracted from algae using a basic solution [9]. The extracted material is then reacted with acid to form alginic acid. The composition of alginate (the ratio of the two uronic acids and their sequential arrangements) varies with the source. Salts of alginic acid with monovalent cations such as sodium alginate are all soluble in water and capable of holding a large amount of water. Alginate has been extensively used as a scaffold for liver, bone, nerve and cartilage engineering [1]. Even though, alginates are non-toxic and biocompatible, using them for biomedical applications has several drawbacks. Alginates are mechanically very weak in their swollen state, therefore it should be blended or modified or copolymerized with other biopolymers before being used as a structural scaffold. More importantly it shows poor cellular adhesion. The chemical structure of sodium alginate is demonstrated in **Figure 1**.

By forming alginate into fibers, novel biomaterials are attainable which can be processed further into woven, non-woven, braided, knitted and many other kinds of composite structures. In the wet-spinning process in which alginate is transformed from powder into a fibrillar-shape, alginate powder is needed to be dissolved in water and stirred properly to form a homogenous solution first. The final properties of wet-spun alginate fibers highly depend on a number of factors, such as chemical structure and molecular weight of the alginate, composition of the coagulation bath, drawing ratio, temperature and feeding rates, etc. The spinning solution is one of the first main considerations in the wet-spinning process which determines the production efficiency. The fiber final performances strongly depends on several parameters including concentration, temperature and pH of the spinning solution [10]. A concentrated sodium alginate solution can be extruded through spinneret holes into a calcium chloride ($CaCl_2$) bath, whereby the high acid content allows alginic acid to undergo spontaneous and mild gelling in the presence of di- or trivalent cations. Thus, it is possible to use a variety of metal ions such as zinc, silver or other bioactive metal ions to precipitate sodium alginate solution during the wet-spinning process as tried previously [11]. Among divalent ions, calcium has found greatest popularity for gel formation of alginate fibers mainly because its salts are cheap, readily accessible and cytocompatible. Since processing takes place in an aqueous solution and in an aqueous coagulation bath at a neutral pH, many bioactive materials, such as drugs and enzymes, can be combined into the alginate fibers, without loss of their bioactivity. On the other hand, calcium alginate fibers have proven to be unstable structures as tissue scaffolds or drug vehicles for in-vivo usages [11, 12].

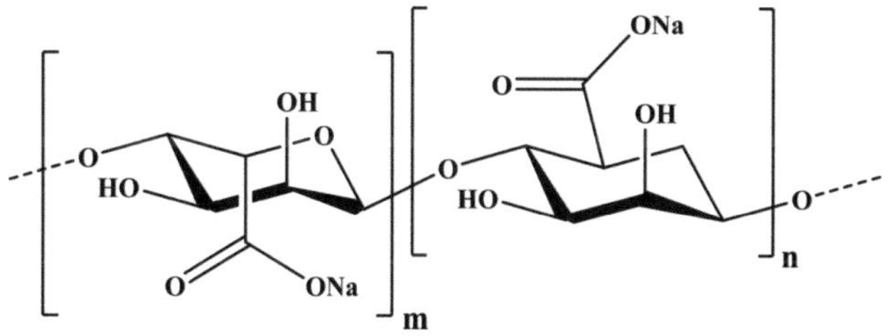

Figure 1. Chemical structure of sodium alginate.

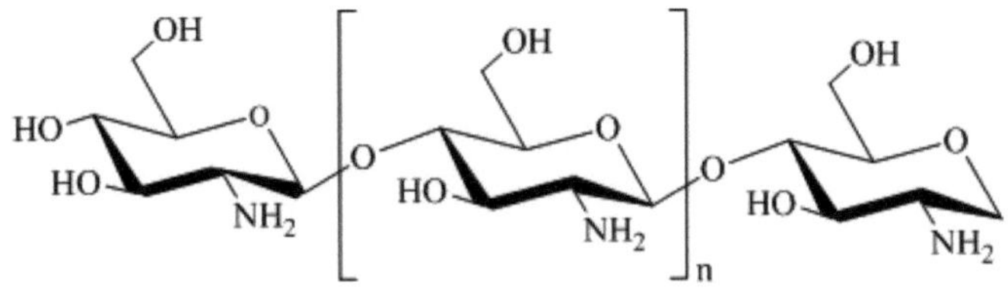

Figure 2. Chitosan chemical structures.

1.4. Chitosan

Chitosan is a semi-crystalline natural polysaccharide [1] with a totally different nature than that of alginate which has recently generated great interest for its potential in clinical and biological applications such as artificial skin, tissue engineering and controlled drug delivery. The cationic polymer chitosan originates from crustacean skeletons [13]. Structurally, chitosan is a semi-synthetically derived aminopolysaccharide which is the N-deacetylated product of chitin, i.e. poly-(1 → 4)-2-amino-2-deoxy-β-D-glucose [1]. Chitosan shows an enhanced hydrophilicity compared to that of chitin which results in a considerable loss of tensile strength in the wet state. This hydrogel is highly reactive due to free amine groups and is readily soluble in weakly acidic solutions resulting in the formation of a cationic polymer of chitosan acetate with a high charge density. These solutions generally have high solution viscosities due to the phenomenon known as the polyelectrolyte effect [14]. Porous chitosan matrix has been used as a scaffold for skin, liver, bone and cartilage, cardiac, corneal and vascular regenerative tissue remodeling. It has also been applied in controlled drug delivery in different shapes such as spheres, films or fibers. The chemical structure of chitosan is shown in **Figure 2**.

Chitosan can be produced in a variety of forms including films, fibers, nanoparticles and microspheres. There have been many attempts by several groups into aqueous basic coagulating baths to produce chitosan fiber [1, 11, 15]. For the purpose of wet-spinning, the chitosan solution is generally extruded into an alkaline solution such as aqueous NaOH as the coagulation bath which forms the fibers. The coagulation rate, which also includes the regeneration of the free amine form of chitosan, is also expected to be influenced by high solution viscosity. Nevertheless, the strong alkaline condition ($pH > 12$) needed to form chitosan-based structures, can limit its utilization for loading most of drugs or bioactive molecules into it.

2. Fabrication methods

2.1. Wet-spinning

The number of fabrication methods for developing three-dimensional structures to be utilized for biological applications have risen due to the inability of two dimensional structures to mimic the extracellular matrix accurately. To design a three-dimensional architecture which imitates the ECM, several parameters such as geometry, mechanical and surface properties as well as biocompatibility are required to be taken into account. Fibers spinning [16] has been used to produce synthetic fibers by first dissolving the desired polymer in a suitable solvent

http://dx.doi.org/10.5772/intechopen.74188

and directly extruding it into a coagulation bath. The bath must contain a liquid which is miscible with the spinning solvent but a non-solvent of the polymer, hence, the solvent is removed from the polymer, leaving a solid fiber which precipitates out of the solution.

Wet-spinning is usually subdivided into three main steps based on different spinning strategies as follows; (a) phase separation, (b) gel separation and (c) liquid crystal spinning. During the phase separation, rapid formation of the fiber structure will occur as a result of polymer solution exposure with the coagulation bath. As the polymer fluid is injected into the non-solvent, the solvent is extracted from the polymer solution causing the polymer to be precipitated in the bath to form a semi-solid fiber. Further solidification into a coagulation bath provides sufficient cohesion and strength for the fiber to be continuously collected when coming out of the coagulation bath. In the second step, the polymer is coagulated due to intermolecular bonds such as ionic cross-linking by a salt or another reacting agent. In the liquid crystal spinning stage, lyotropic crystalline solution provides sufficient alignment and cohesiveness to form a solid crystalline phase for fibers. A schematic of wet-spinning is shown in **Figure 3**. Although coaxial electrospinning was used by many researchers to produce coaxial fibers, only a few reports which appeared in the literature reported the successful fabrication of hybrid fibers *via* coaxial wet-spinning methodology to the knowledge of author. This might be due to the complexity of this method because of plurality of parameters involved in the successful formation of a core-sheath structure inside a coagulation bath. In the first instance, the fabrication of wet-spun coaxial fibers appears a straightforward task; two different components are injected through a coaxial spinneret at once into a proper coagulation bath to form a coaxial structure. However, this simple approach presents several challenges. Many parameters are needed to be controlled and regulated in order to hold both components in a coaxial structure. Among those, the solution properties are known as the key factors affecting the spinning process including mainly the material concentrations, viscosities, surface charges, surface tensions, polymer natures and functionalities and so on. However, less important considerations such as systematic variables (injection rates (V_i), core to sheath injection rates ratio, take-up velocity (V_t), coagulation bath constituents, drawing velocity (V_d), spinneret specialties, post-treatment processes, etc.) and ambient conditions (temperature, post-

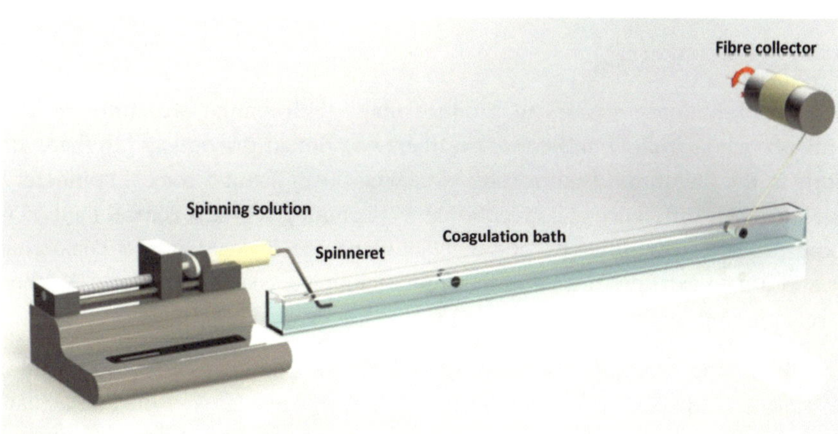

Figure 3. Schematic of a Lab scale wet-spinning line.

spinning conditions) could be also influencing the process as well as final fiber properties significantly. Up to date, different examples of coaxial fibers have been reported using a number of materials such as conducting polymers, metals, natural polymers and carbon-based components *via* various production methods. Most efforts have been focussed on approaches based on electrospinning to date to produce coaxial structures. Some of the main techniques to produce coaxial fibers and yarns are described in the following sections. However, among earlier efforts to produce wet-spun coaxial fibers, few methods could be found sharing similar procedures to that of coaxial wet-spinning described below [16].

2.2. Production of hollow fibers

Many different types of semipermeable hollow fibers have been prepared by means of the melt, dry or wet-spinning techniques [1]. The procedure of achieving a hollow structure is quite similar to that of coaxial wet-spinning in which the sheath spinning solution was delivered from a chamber to the external spinneret nozzle through injection, whereas the core fluid is usually substituted simply with pressurized water or the coagulant fluid injected to the central nozzle. These fibers are being applied in various purposes such as gas separation, ultrafiltration, reverse osmosis as well as many biological applications including drug delivery, dialysis and tissue engineering.

Delivery systems with diverse release profiles spanning from a few days to several months have been achieved by encapsulation of biological molecules as well as drug reservoirs into the core of hollow fibers or chemically cross-linking or adsorbing therapeutics to the surfaces of fibers. With regard to the encapsulation of drugs within wet-spun filaments, a critical issue is obtaining the appropriate release characteristics and mechanical integrity for specific cell type/tissue architecture. The fabrication of the hollow PLGA fiber has been done as a controlled drug release system [15]. A method also *has* been described for encapsulation of human hepatocellular carcinoma cells in wet-spun chitosan-alginate microfibres [1]. In addition to those mentioned, many other research groups reported on using a hollow fiber structure for delivering biomolecules [1, 11, 15].

2.3. Coaxial wet-spinning

Despite those preliminary studies to produce fibers with similar structures to that of the coaxial fibers such as hollow and core-skin fibers mentioned previously [3], there are only a few reports in the literature of fabrication of coaxial fibers using a coaxial spinneret for wet-spinning. The major difference of conventional wet-spinning with the coaxial method is that in the coaxial process, two different polymer solutions are injected into a coaxial spinneret together and are co-extruded into a bath while retaining a coaxial structure. A schematic of coaxial wet-spinning setup is shown in **Figure 4**.

Coaxial wet-spinning produces hollow or core-shell fibers that can be used for quite a lot of purposes such as controlled release applications, electronic textiles, sensors and actuators [12]. As a matter of fact, for the successful production of coaxial fibers, several parameters are needed to be controlled and regulated. Thus, development of a simple yet effective wet-

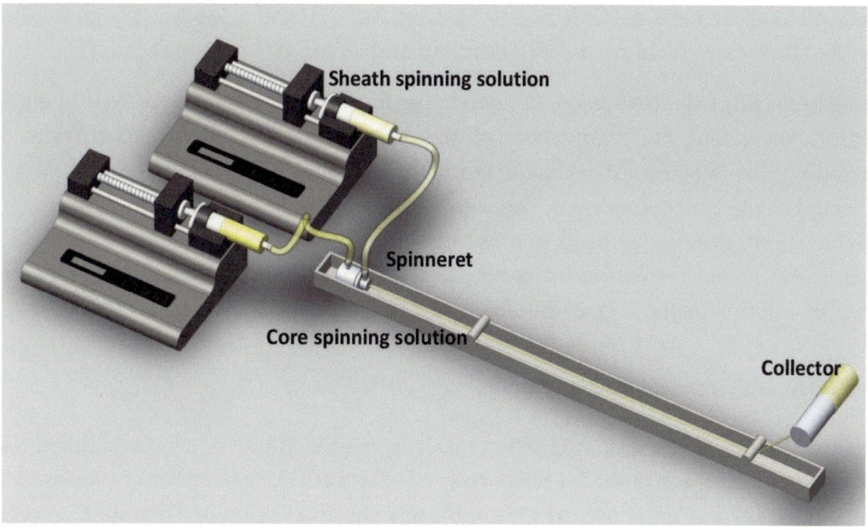

Figure 4. Coaxial wet-spinning setup.

spinning approach for direct preparation of sheath-protected fiber electrodes still remains a challenge. Among those, solution properties are of great importance such as material concentrations, viscosities, surface charges, surface tensions, polymer natures and functionalities etc. However, process parameters could also influence the process as well as the final fiber properties significantly. G. Park and his co-workers were successful to spin CNT/Poly (vinyl alcohol) fibers with a sheath-core structure *via* wet-spinning [1]. Recently, Liang Kou *et al.* have also reported on the production of CMC/wrapped graphene/CNT coaxial fibers for supercapacitor applications [15].

3. Materials and methods

3.1. Wet-spinning of hydrogels fibers

Wet-spun chitosan fibers were produced in a coagulation bath consisting of 1 M sodium hydroxide (NaOH) using a rotary wet-spinning system. Uniform alginate fibers were spun in a 2% $CaCl_2$ coagulation bath. Core-sheath fibers of chitosan-alginate (Chit-Alg) were successfully spun using a coaxial spinneret. The chitosan spinning solution (with different amounts of $CaCl_2$) was injected as the core component and extruded through the center nozzle into the coagulation bath of calcium chloride [11]. Simultaneously, alginate was injected as sheath of the fiber, providing an outer casing for the core, by injection through port A. In this method, by using a blend of chitosan with various percentages of calcium chloride, it is possible that the alginate sheath can be coagulated from the inner chitosan core, while also creating the opportunity to react chitosan with sodium alginate at a much faster rate [1]. The setup is shown in **Figure 4**, previously. Therefore, as mentioned earlier chitosan solutions including 0.5, 1 and 2% ($w\ v^{-1}$) $CaCl_2$ were prepared for the core component of the fibers and alginate sheath. The samples are named here as Chit-Alg (0.5), Chit-Alg (1) and Chit-Alg (2). Solutions were delivered at flow rates of 14 mL h^{-1} for chitosan and 25 mL h^{-1} for the sheath [11].

Toluidine blue (TB) was used as an indicative dye incorporated into the coaxial fibers to track the release experiment. For the purpose of fiber preparation for release experiments, the dye was mixed with chitosan solution before spinning with the concentration of 0.1% (w v^{-1}) and then injected as the core component. These solutions were then spun into the same coagulation baths which were previously used to make pristine fibers. Coaxial fibers containing TB were also fabricated using the method mentioned previously with the small difference of using chitosan/TB solution as the core component [11].

3.2. Toluidine blue release measurement

The release kinetics of the prepared fibers for drug release applications was studied using TB as a model dye introduced into the fibers over a 5-day period. The amount of released TB was determined *via* UV-vis spectroscopy by monitoring the absorption of TB at its λ_{max} 630 nm in simulated body fluid (SBF). To construct an absorbance calibration curve for sample analysis using a Shimadzu UV 1601 spectrophotometer, UV-vis spectra of SBF solutions containing TB with different concentrations were recorded between 200 nm and 1100 nm. Approximately 5 cm of each dried sample (in triplicate) was placed in a 2 mL Eppendorf tube and 1 mL of SBF was added into it. The release medium was taken by micro-pipette at specific time points over 5 days and replaced with the same volume of fresh SBF solution to maintain the total volume constant. The percentage of released TB (%) was plotted versus time [11].

4. Results and discussion

Initial investigations were aimed at determining the spinnability and the physical characteristics of potential dopes to ensure they were in the range required for fiber formation.

4.1. Spinnability vs. concentration

Spinnability can be defined as the ability of a material of being suitable for spinning or the capability of being spun. In the context of wet-spinning, spinnability could be referred to the ability of a solution to form fibrillar arrangements *via* injection into a non-solvent medium which makes it precipitate, so-called a coagulation bath [16]. The spinnability of a polymer solution depends on many parameters, including the rheological properties of a solution, size of nozzle, shear rate applied during injection through spinneret and mass transfer rate difference between the extruded solution and non-solvent. Several types of either suitable solvent or coagulant could be employed depending on the chemical structure of material. Often an upper and lower limit for polymer concentration during wet-spinning is considered facilitates spinning continuous length of fibers. The capillary break-up is widely understood as a surface-tension induced break-up of filaments into drops can occur at low concentrations of the polymer solution which determines the lower limit of spinnability. Spinnable concentrations have been reported for chitosan varying from 2 to 15% (w v^{-1}). Here, we found that 2–5% (w v^{-1}) is the appropriate concentration range enabling wet-spinning of MMW chitosan into a coagulation bath of 1 M NaOH. Observations also indicated that aqueous alginate solutions at a concentration of below 2% (w v^{-1}) would not generate a continuous fibrous structure *via* wet-spinning; increasing the alginate concentration from 2 to 4% (w v^{-1}), the solution became highly spinnable [1]. Then again, at concentrations above 4% (w v^{-1}), the solution became highly viscous which imped continuous flow through the needle, rendering the solution unspinnable. A concentration of 3% (w v^{-1}) has been thus selected for both gel precursors due to the ease of spinnability, together with maintaining the suitable mechanical properties for coaxial wet-spinning [1].

4.2. Rheology

Viscosity is considered in the selection of suitable concentrations of chitosan and alginate solutions for fiber spinning. For coaxial spinning matching viscosities of the two components is also a consideration [1]. **Figure 5** shows changes in viscosity versus shear rate was determined from aqueous solutions of chitosan at 3% (w v^{-1}) and alginate at 3% (w v^{-1}). Spinning solutions of 3% (w v^{-1}) chitosan resulted in a solution with a viscosity of 6.4 Pa·s. The viscosity of 3% (w v^{-1}) sodium alginate solution was approximately 8.5 Pa·s. The viscosities of the two solutions became closer as the shear rate increased. Under shear, hydrogel chains are in a less expanded conformation and become less entangled causing the viscosity to drop. At the time of spinning, chitosan is injected with rate of 14 mL h^{-1} while is 25 mL h^{-1} for the alginate solution. The shear rates calculated to be about ~97 s^{-1} for alginate and ~75 s^{-1} for chitosan

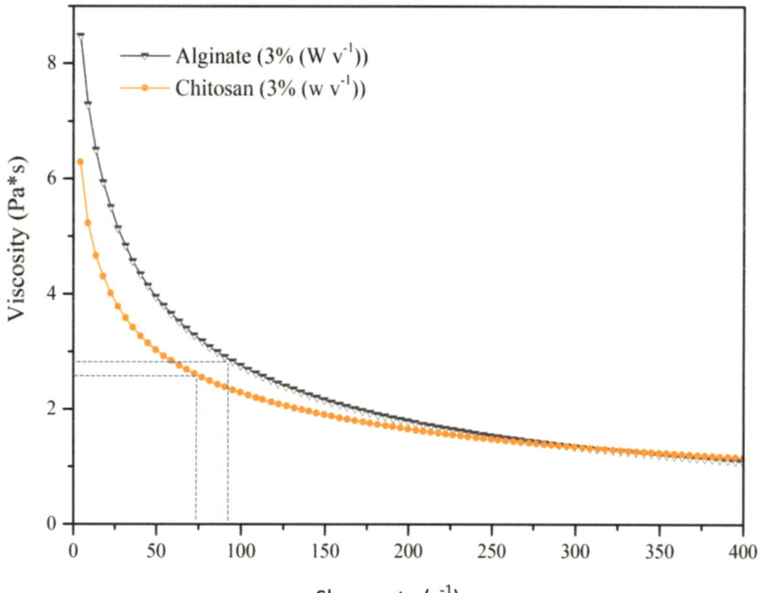

Figure 5. Viscosities of spinning solutions of chitosan and sodium alginate [11]. Reproduced with permission. 158 Copyright 2015, Wiley-VCH.

solutions which resulted in a viscosity of ~2.5 Pa·s for chitosan and ~2.8 Pa·s for alginate solutions. These outcomes seem to be ideal for coaxial spinning [1].

4.3. Continuous spinning of coaxial fibers

To produce continuous uniform fibers, chitosan with injection rate of 14 mL h^{-1} and alginate solution with rate of 25 mL h^{-1} have been simultaneously injected into the 2% (w v^{-1}) aqueous $CaCl_2$ coagulation bath through the ports built in the coaxial spinneret [11]. Using this method, an unlimited length of fibers could be obtained which was collected using a collector as shown in **Figure 6**. It is worth mention that the preparation of coaxial fibers without incorporating a certain amount of $CaCl_2$ did not turn out to be successful when tried.

The woven structure of Chit-Alg fibers (containing TB) in dry and wet state were shown in **Figure 7 (a)** and **(b)**, respectively. This capability provides the potential for these structures to be utilized as tissue scaffolds and drug delivery vehicle applications [11].

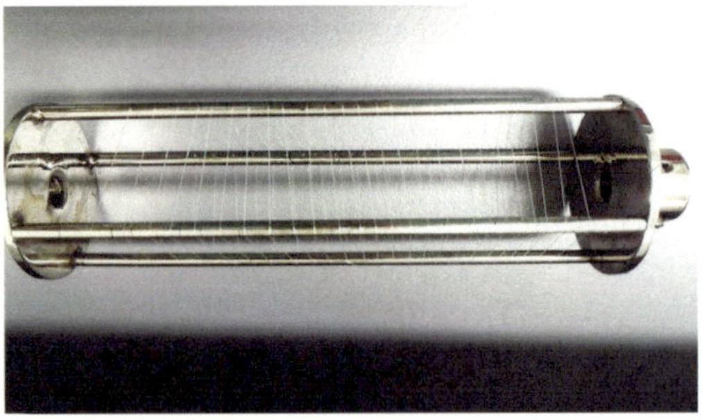

Figure 6. The capability of producing unlimited length of coaxial Chit-Alg (1) fibers as shown onto a collector. [11] Reproduced with permission. 158 Copyright 2015, Wiley-VCH.

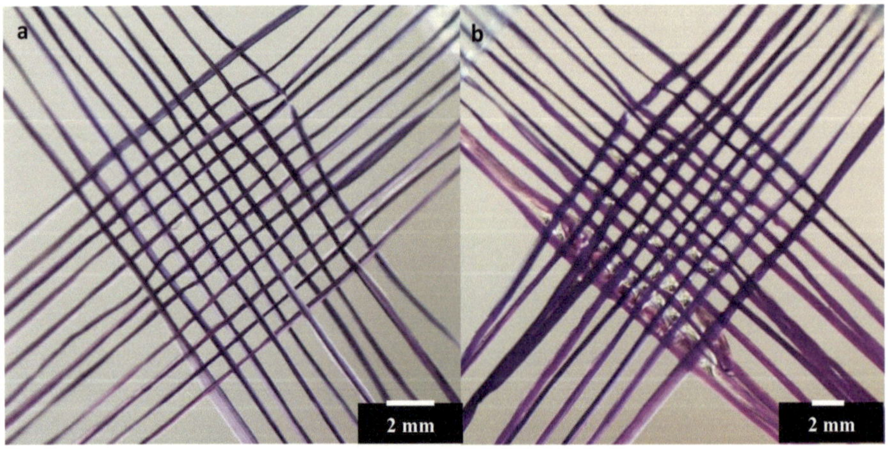

Figure 7. The photographs of scaffold structure woven by coaxial fibers; imaged in (a) dry state and (b) wet state. [11] Reproduced with permission. 158 Copyright 2015, Wiley-VCH.

http://dx.doi.org/10.5772/intechopen.74188

4.4. Morphology of As-prepared fibers

The stereomicroscope images of wet-spun chitosan, alginate and core-sheath Chit/Alg fibers are shown in **Figure 8** in wet and dry-states.

As can be seen in **Figure 8 (a)** and **(b)**, the surface of the chitosan fiber seemed to be very smooth and soft, while some wrinkles can be noticed spreading on the surface of the alginate fiber which will be increased during the fiber drying process [1]. Moreover, one can see that the

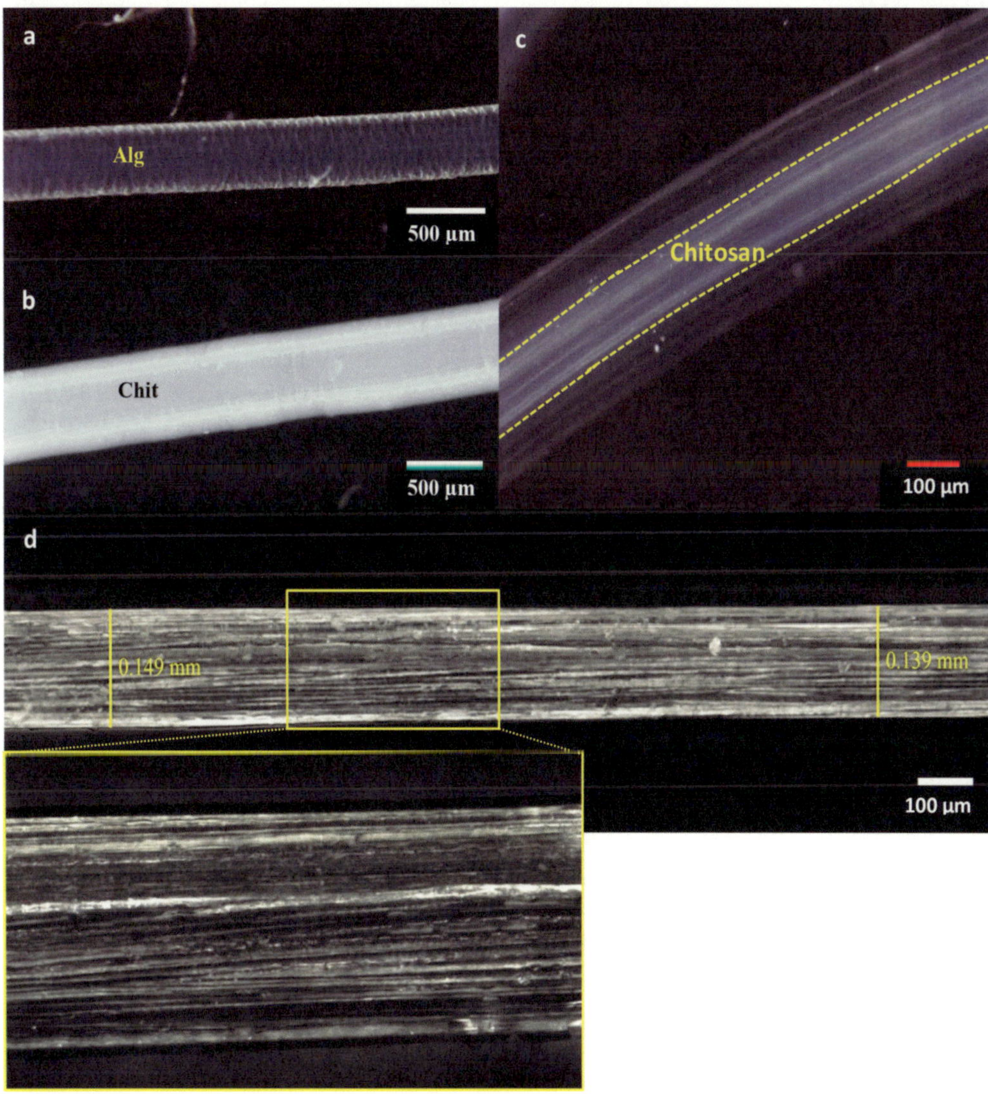

Figure 8. Stereomicroscope images of side view of wet (a) alginate, (b) chitosan, (c) coaxial Chit-Alg (1) and (b) dry Chit-Alg (1) fiber. [11] Reproduced with permission. 158 Copyright 2015, Wiley-VCH.

core-sheath fibers are straight and smooth with a core loaded in the center of fibers. They have a uniform structure and diameter of ca. 220 μm and 136 μm for the sheath and core when wet, respectively (**Figure 8(c)**). However, the dried fibers hold the thickness of ~140 ± 10 μm (This average value was calculated after measuring the diameter under the stereomicroscope 10 times). In addition, some lines or longitudinal indentations can be observed running parallel with the fibers on totally dried core-sheath fiber as shown in **Figure 8 (d)**. The chitosan core is ~90 μm which is surrounded by a thin layer of alginate sheath of ~8–12 μm. Still, the thicknesses of core and sheath materials are adjustable by changing solution feed rates and the drawing ratio (data, variables *vs.* dimensions). Considering two selected collection rates at angular velocities of 20 and 60 rpm (with the assumption of keeping the injection rates constant), it is possible to measure draw ratio upon increasing the collection rate from 20 to 60 rpm while having a constant collector [1].

The fiber diameter decreased as the drawing ratio increased. In general, the molecular orientation of fiber materials obtained through the drawing process governs their properties, particularly the mechanical properties. In addition, the thickness of the sheath would have a direct relationship with increasing its injection rate; the thickness of alginate increased as the shear rate increased while the feeding rate of core component was kept constant. However, while the application of shear is essential in obtaining orientation in the fiber, high shear rates develop beaded non-uniform fiber in the coagulation bath as a result of die swell (swelling of the free jet of solution upon injection from spinneret) and skin formation. Die swell occurs as a consequence of polymer relaxation due to its low entropy conformation after shear is applied during extrusion through the spinneret, where polymer molecules are oriented by the flow. The diameter of the jet then decreases as a result of drawing along the spinning path. A hard skin is also formed on the surface of the filament which results in the rate the jet diameter decreases. When the shear rate of chitosan increased to 20 mL h^{-1}, formation of the coaxial structure did not turn out to be successful. It seemed that the sheath components were not thick enough to hold the core material in place.

SEM images of cross-sections and the surfaces of solid and core-sheath fibers are illustrated in **Figure 9 (a–e)**. They give valuable information about the morphology of the two polymers. Before imaging fibers were immersed in SBF and imaged with SEM in an attempt to capture structural information in the "wet state", since that is how they would be used in future possible applications [1].

Cross-sections of solid fibers fabricated showed the cylindrical shaped form of the hydrated chitosan and alginate solid fibers (**Figure 9 (a)** and **(b)**), respectively. Alginate fibers appeared to be permeable and spongy, while the cross-section of chitosan fibers appeared to be denser. In contrast, cross-sections of the coaxial fibers reveal slightly irregular, oval shaped fibers with a distinct separation between chitosan in the core and the outer alginate sheath as is indicated in **Figure 9(c)**. In addition, both polymers showed an extensive porous structure in the coaxial structure. On the cross-section of chitosan, regular crystalline structures can be seen which are probably due to the presence of calcium chloride inside the core (**Figure 9 (d)**, while alginate has a honeycomb structure (**Figure 9 (e)**). It is evident that the fiber is composed of two distinct areas of chitosan and alginate [1].

http://dx.doi.org/10.5772/intechopen.74188

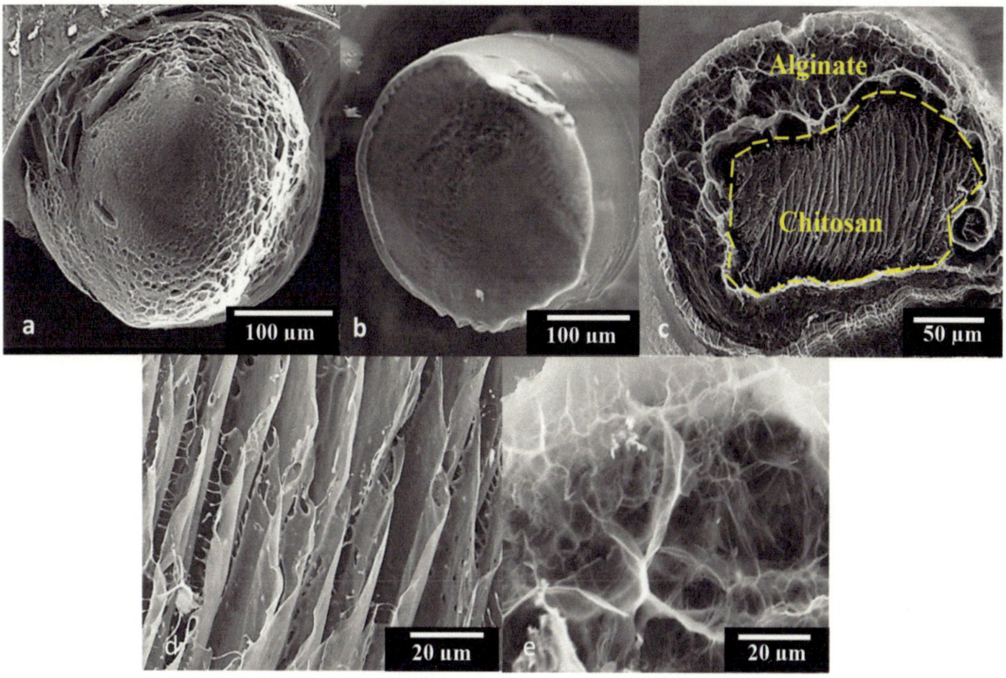

Figure 9. LV-SEM images of hydrated as-prepared (a) alginate, (b) chitosan and (c) Chit-Alg (1) cross-section in SBF, (d) chitosan core arrangement in cross-section (e) alginate sheath construction in the cross-section. [11] Reproduced with permission. 158 Copyright 2015, Wiley-VCH.

4.5. Mechanical properties of as-prepared fibers

The mechanical properties of alginate, chitosan and Chit-Alg coaxial fibers employing different concentrations of calcium chloride in chitosan core spinning dope are depicted in **Figure 10**. Ultimate stresses (MPa), ultimate strains (%), Young's moduli (MPa) and swelling ratios (%) were measured for alginate, chitosan, Chit-Alg (0.5), Chit-Alg (1) and Chit-Alg (2) fibers, respectively [11].

Mechanical analysis results revealed that with addition of more calcium chloride to the core-dope, the Young's modulus decreased. Increasing the amount of calcium chloride into fiber core will probably cause agglomerations which can lead to phase separation. Thus, there would be an upper threshold for the amount of $CaCl_2$ in the core at which the optimum mechanical parameters could be achieved. As a result, the mechanical properties of as-prepared fibers such as Young's modulus and ultimate stress were decreased by addition of more than 1% (w v^{-1}) $CaCl_2$.

The results, which are presented in **Table 1**, also confirmed the reinforcing role played by the chitosan core in coaxial Chit-Alg fibers. Young's modulus was measured to be *ca.* 1.7 and 6.6 MPa for alginate and chitosan solid fibers, respectively. It has been also revealed that the fibers which contain 1% (w v^{-1}) $CaCl_2$ resulted in the highest mechanical results due to their modulus and ultimate stress compared to other coaxial fibers [11].

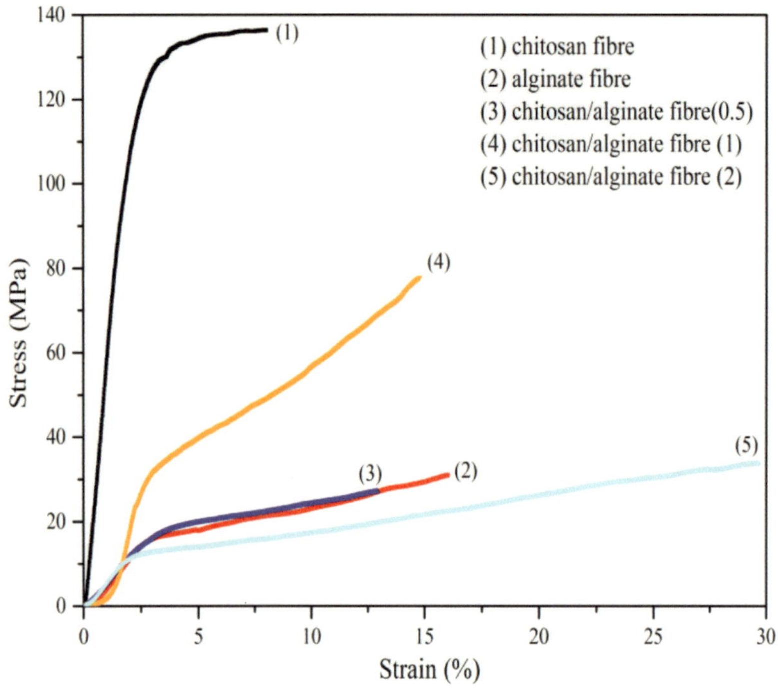

Figure 10. Stress-strain curves obtained from tensile tests of alginate sinzgle and chitosan/alginate coaxial fibers using different $CaCl_2$ concentrations. [11] Reproduced with permission. 158 Copyright 2015, Wiley-VCH.

Sample	Breaking stress (MPa)	Strain at break (%)	Young's modulus (GPa)	Initial swelling ratio (%)
Alginate fiber	~31 ± 5	~26 ± 3	1.6 ± 0.15	(Non-measurable)
Chitosan fiber	~146 ± 30	~19 ± 5.2	6.6 ± 0.8	~90
Chit-Alg (0.5)	~30 ± 5	~22 ± 8	0.6 ± 0.1	~360
Chit-Alg (1)	~80 ± 10	~14 ± 3	1.9 ± 0.2	~540
Chit-Alg (2)	~28 ± 5	~37 ± 5	0.55 ± 0.1	~385

Table 1. Comparison of mechanical properties of solid and coaxial biofibers [11]. Reproduced with permission. 158 Copyright 2015, Wiley-VCH.

4.6. Swelling properties in SBF

The swelling properties of the fibers were determined in SBF medium over a period of 48 hrs. Fiber diameters were measured at different time intervals. Results are shown in **Figure 11** and listed in **Table 1**. Solid fibers have shown quite different degrees of swelling; while chitosan fibers showed only 90% calcium alginate fiber, the swelling of alginate fibers occur quite fast up to high ratios (until the fiber lose its fibrillar shape completely which make it almost impossible to be measured). This phenomenon is mostly due to the ionic exchange between the divalent cations and sodium in the environment.

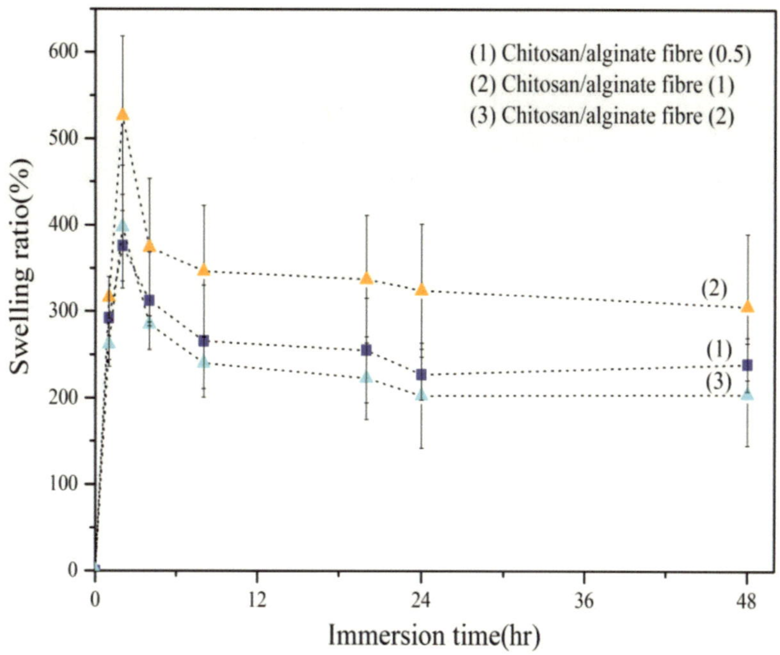

Figure 11. Swelling properties of coaxial wet-spun fibers in SBF as a function of the immersion time. [11] Reproduced with permission. 158 copyright 2015, Wiley-VCH.

It can be seen in **Figure 11** that coaxial fibers containing 0.5% (w v^{-1}) calcium chloride have shown the least amount of initial swelling, while fibers with 1% (w v^{-1}) calcium chloride in the core demonstrated the highest degree of swelling [1]. It seems that two simultaneous events are occurring by increasing the calcium chloride content in the core (from 0.5 to 2% (w v^{-1})). Increasing the number of ionic groups (Ca^{2+}) in hydrogels is known to increase their swelling capacity [17]. This is mainly due to the simultaneous increase of the number of counterions inside the gel, which produces an additional osmotic pressure that swells the gel as described in Flory theory previously. Therefore, by adding more calcium chloride to the chitosan solution the degree of swelling increases. On the other hand, increasing the amount of Ca^{2+} ions also results in an increase in the ion exchange process within the sodium alginate. In fact, the ratio of calcium will increase in the alginate. The increase of the cross-linking agent concentration leads to the formation of a hydrogel with a greater 3D network density and so results in sheaths which show less swelling [1].

4.7. *In-vitro* release measurement

The calibration curve was determined by monitoring the absorption of TB at its λ_{max} (630 nm) in SBF with various concentrations of TB using UV-vis spectroscopy. The ability of the drug to release from the polymer matrix depend on a number of factors such as the solubility of the drug in the polymer matrix, the solubility of the drug in the medium, swelling and solubility of the polymer matrix in the medium and the diffusion of the drug from the polymer matrix to the medium [18]. The release profiles of TB from dye loaded coaxial fibers in SBF for up to 5 days were plotted *vs.* time and are demonstrated in **Figure 12**.

The whole release time period varied for different types of fibers including alginate, chitosan and the core-sheath fiber depending on the period over which they could resist the media before their structure fell apart [1]. As noted previously, calcium alginate could be easily degraded when used for *in-vivo* applications due to the ionic exchange between the divalent cations and sodium in the which are present in the body environment [11, 18]. Therefore, it is believed that the release of TB observed from alginate fibers was mainly due to the degradation of alginate fibers. On the other hand, wet-spinning of chitosan fiber is needed to be done in basic coagulation bath which is not an appropriate condition for most of loaded drugs. Coaxial fibers indicated a controlled manner of release more or less like chitosan fibers. However, with the help of coaxial spinning, their fabrication process *via* wet-spinning is performed in a neutral coagulation bath. These results provide the suitable condition to load any types of drugs into the wet-spun fibers for drug delivery applications. As can be seen in **Figure 12**, the coaxial fibers showed similar release behavior to that of the chitosan fibers. However, they could withstand the media for a shorter period of time without losing the initial structure.

In the initial period of 2 h, a fast release of TB from alginate fibers is observed at which more than 70% of TB is released. Either chitosan or Chit/Alg coaxial fibers showed approximately 30% burst release of TB followed by a sustained release within over 5 days. While alginate fiber could not withstand the media for more than 4 days, *ca.* 42 and 50% of the TB is released from chitosan and Chit/Alg fibers, respectively. **Figure 12** shows a good sustained-release profile of

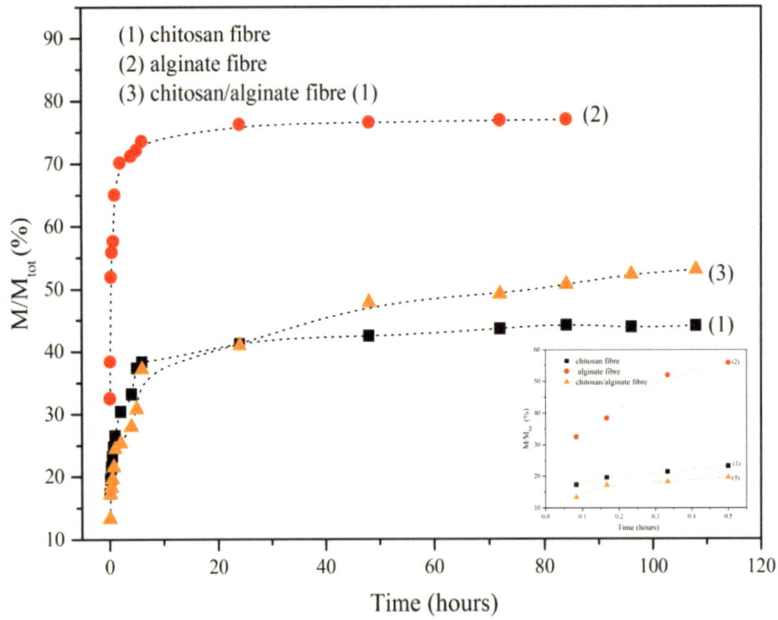

Figure 12. Time dependent TB releasing behavior of chitosan, alginate and Chit/Alg hydrogel fibers in SBF at 37°C. Inset; burst release of coaxial fibers in the first 30 min. [11] Reproduced with permission. 158 Copyright 2015, Wiley-VCH.

TB from coaxial fibers. TB is a hydrophilic molecule with a greater solubility in aqueous environment, so its drug diffusion rate through the polymeric matrix is highly dependent on the swelling of the polymeric fiber. Thus, according to the swelling ratio results, it is expected to obtain much faster release from alginate fibers than those of either chitosan or coaxial fibers [11].

5. Conclusion

The development and fabrication of hydrogels fibers has been carried out to evaluate their performance for drug delivery systems. The production of coaxial hydrogels fibers were successfully developed for the first time using a wet-spinning method. The morphological, mechanical, thermal and swelling properties of these fibers are discussed [1]. Enhanced mechanical properties of 260% in ultimate stress and more than 300% in the Young's modulus were observed by incorporating 1% (w v^{-1}) $CaCl_2$ into the chitosan core. SEM micrographs of the cross-section of chitosan-alginate fibers clearly show the cylinder shaped monofilament form of the chitosan fiber covered with alginate. These biofibers as delivery platforms have demonstrated great potentials toward advancing current drug delivery systems. Hybrid Chit/Alg fibers could likely be promising as a novel kind of 3D bioscaffolds in drug release studies or tissue engineering [1].

Acknowledgements

This work was supported by funding from the Australian Research Council under Discovery Early Career Researcher award (Javad Foroughi DE12010517). The authors would also like to thank Saber Mostafavian for 3D design of the wet-spinning process.

Conflict of interest

The authors declare that there is no conflict of interest; this chapter is wholly our own work unless otherwise referenced or acknowledged. The document has not been submitted for publication at any other publishing organization.

Acronyms and Abbreviations

3D	three-dimensional
ECM	extracellular matrix
Chit	chitosan
Alg	alginate

$CaCl_2$ calcium chloride

V_i injection rate

TB toluidine blue

SBF simulated body fluid

Author details

Javad Foroughi*, Azadeh Mirabedini and Holly Warren

*Address all correspondence to: foroughi@uow.edu.au

Intelligent Polymer Research Institute, University of Wollongong, NSW, Australia

References

[1] Mirabedini A. Developing Novel Spinning Methods to Fabricate Continuous Multifunctional Fibres for Bioapplications. Doctor of Philosophy thesis. Australia: University of Wollongong; 2017. http://ro.uow.edu.au/theses1/6

[2] Elahi F, Lu W, Guoping G, Khan F. Core-shell fibers for biomedical applications-a review. Journal of Bioengineering and Biomedical Science. 2013;**3**:1-14

[3] Abbas AA, Lee SY, Selvaratnam L, Yusof N, Kamarul T. Porous PVA-chitosan based hydrogel as an extracellular matrix scaffold for cartilage regeneration. European Cells & Materials. 2008;**16**:50-51

[4] Breyner NM, Zonari AA, Carvalho JL, Gomide VS, Gomes D, Góes AM. Biomaterials and Biotechnology Schemes Utilizing TiO_2 Nanotube Array. Biomaterials Science and Engineering. Brazil: InTech Published, Institute of Biologic Science, Department of Biochemistry and Immunology; 2011. pp. 211-226

[5] Martins A, Reis RL, Neves NM. Electrospun nanostructured scaffolds for tissue engineering applications. Nanomedicine. 2007;**2**:929-942

[6] Zhang T, Wan LQ, Xiong Z, Marsano A, Maidhof R, Park M, Yan Y, Vunjak-novakovic G. Vunjak-novakovic, Channelled scaffolds for engineering myocardium with mechanical stimulation. Journal of Tissue Engineering and Regenerative Medicine. 2012;**6**:748-756

[7] Draget KI, Smidsrùd PO, Skjåk-brñk PG. Polysccharides and Polyamides in the Food Industry. Properties, Production and Patents. KGaA, Weinheim: Wiley-VCH Verlag GmbH & Co; 2005. pp. 1-30

[8] Zhu C, Fan D, Duan Z, Xue W, Shang L, Chen F, Luo Y. Initial investigation of novel human-like collagen/chitosan scaffold for vascular tissue engineering. Journal of Biomedical Materials Research. Part A. 2009;**89**:829-840

[9] Wang L, Khor E, Wee A, Lim LY. Chitosan-alginate PEC membrane as a wound dressing: Assessment of incisional wound healing. Journal of Biomedical Materials Research. 2002;**63**: 610-618

[10] Kuo CK, Ma PX. Ionically crosslinked alginate hydrogels as scaffolds for tissue engineering: part 1. Structure, gelation rate and mechanical properties. Biomaterials. 2001;**22**:511-521

[11] Mirabedini A, Foroughi J, Romeo T, Wallace GG. Development and characterization of novel hybrid hydrogel fibers. Macromolecular Materials and Engineering. 2015;**300**:1217-1225

[12] Foroughi J, Spinks GM, Wallace GG. A reactive wet spinning approach to polypyrrole fibres. Journal of Materials Chemistry. 2011;**21**(17):6421-6426

[13] Bhavan M, Nagar G. Chitosan–sodium alginate nanocomposites blended with cloisite 30b as a novel drug delivery system for anticancer drug curcumin. International Journal of Applied Biology and Pharmaceutical Technology. 2011;**2**:402-411

[14] Hussain A, Collins G, Yip D, Cho CH. Functional 3-D cardiac co-culture model using bioactive chitosan nanofiber scaffolds. Biotechnology and Bioengineering. 2012;**110**:1-11

[15] Mirabedini A, Foroughi J, Thompson B, Wallace GG. Fabrication of Coaxial Wet-Spun Graphene–Chitosan Biofibers. Advanced Engineering Materials. 2015

[16] Mirabedini A, Foroughi J, Wallace GG. Developments in conducting polymer fibres: From established spinning methods toward advanced applications. RSC Advances. 2016;**6**(50):44687-44716

[17] Okay O, Durmaz S. Charge density dependence of elastic modulus of strong polyelectrolyte hydrogels. Polymer Journal. 2002;**43**:1215-1221

[18] Wade SJ et al. The potential role of gut microbiota in pancreatic disease: A systematic review. Pancreatology. **17**(5):795-804

Chapter 7

Obtaining Hydrogels based on PVP/PVAL/Chitosan Containing Pseudoboehmite Nanoparticles for Application in Drugs

Leila Figueiredo de Miranda,
Kátia Lucia Gonçalves Cunha,
Isabella Tereza Ferro Barbosa,
Terezinha Jocelen Masson and
Antonio Hortêncio Munhoz Junior

Additional information is available at the end of the chapter

http://dx.doi.org/10.5772/intechopen.72007

Abstract

People with skin lesions caused by burns, ulcerations and other complications, independent of degree and extension of the problem, has induced the search for methods and materials to optimize the process of tissue repair in matter of time and quality. Thus, materials made by synthetic polymers have been used and improved due to overwhelming demand. The efficacy of dressings and bandage depends on a variety of factors such as biocompatibility, composition uniformity, low cost, long validity, flexibility, and so on. In this chapter, hydrophilic membranes based on polyvinylpyrrolidone-PVP/poly(vinyl alcohol)-PVAl and chitosan containing nanoparticles of pseudoboehmite for use in pharmaceuticals were developed and studied. The hydrogels were obtained by ionizing radiation in electron-beam accelerator at a dose of 25 kGy and characterized by mechanical, thermal and physicochemical tests. Pseudoboehmite nanoparticles were obtained from aluminum nitrate by a sol–gel process. The characterization of the hydrogels was done by various tests such as tensile, swelling, thermal analysis, sol-gel fraction and dynamic mechanical analysis. The results show that the presence of PVAl hydrophilic membrane causes lower degree of swelling, greater attraction and greater resistance to elongation at break in tension, although significantly lower fraction of gel membranes contains only agar and PVP. It was verified that the presence of chitosan nanoparticles and pseudoboehmite promotes a decrease in the formation of cross-links during irradiation of hydrophilic membranes.

Keywords: PVP, PVAl, chitosan, pseudoboehmite, nanotechnology, hydrophilic membranes

1. Introduction

Hydrophilic membranes based on hydrogels can be defined as a polymeric material, which is insoluble in water, can absorb it and retain a significant fraction in its structure [1]. The material that forms the membranes with these characteristics is composed of two water-insoluble cross-linked hydrophilic polymer systems, due to the existence of a three-dimensional network connecting their chains, and is perceived that water was retained in its structure [2–4].

Hydrogels can be expanded by water and ion absorption, above the equilibrium state and retain their original shape and mechanical properties. In addition, they present permeability to biologically active substances with low molecular weight, being used in dressings directly on contact with the skin and can be used as dressing or bandage in burn wounds, vascular prostheses, artificial cartilage membranes for hemodialysis, among other applications [5, 6].

Hydrogels obtained by ionizing radiation, having as precursors poly(n-vinyl-2-pyrrolidone) (PVP), agar and plasticizing agents such as polyethylene glycol that have been used in pharmaceuticals and have advantages over the previously obtained methods, they eliminate the sterilization phase, as this is still being obtained in the process of ionizing radiation [1].

Due to its properties, PVAl is one of the synthetic polymers that are used as a biomaterial, being of great importance in the industry, mainly in cosmetics, where it is used as an additive, providing texture to the products. Its main function is that of a plasticizer. Its use in the synthesis of hydrophilic membranes presents some restrictions when used alone, because the obtained membrane presents low elasticity and rigidity. To improve these properties, it is used in conjunction with other polymers [7, 8].

Chitosan is a natural polymer that can be obtained through the process of deacetylation of chitin, a polysaccharide of great abundance in nature, and present a similar chain to cellulose. Among its main characteristics, of great importance and industrial interest, it is atoxic and has an easy way to create gel [9].

Chitosan nanocomposites have been used in the cotton coating for application in dressings and bandage with the purpose of increasing the absorption of the exudate as well as improving the antibacterial activity [10].

Chitosan and PVAl are biocompatible polymers being in the composition of several hydrogels used today [11, 12]. Hydrophilic membranes based on PVAl and chitosan have been synthesized because the presence of chitosan improves the mechanical, hydrophilic and antibacterial properties of membranes obtained from PVAl [13, 14]. These membranes modified by the presence of nanoclays for use in dressings have been obtained with good mechanical and absorption properties [15, 16].

Hydrophilic membranes obtained from polyvinyl alcohol, poly(N-vinylpyrrolidone) and antibiotic containing chitosan have been synthesized by Yu et al. [17].

The pseudoboehmite obtained by the sol-gel process is a ceramic nanoparticle with high surface area, bioinert, which can be used in drug delivery systems [18]. Pseudoboehmite has the same structure as the boehmite (ɣ-ALOOH). It has an orthorhombic structure and presents two layers of octahedral oxygen partially filled with aluminum cations [19]. Through the sol:

gel process, it is possible to obtain nanoparticles of pseudoboehmite, having as precursors aluminum nitrate and ammonium hydroxide [20].

The sol-gel process consists of synthesis of materials, caused by the transition of a sol system, dispersion of colloidal particles (dimension from 1 to 100 mm) in a fluid; in a gel system, rigid structure system of colloidal particles or chains polymerization by immobilizing the liquid phase [21].

Pseudoboehmites obtained by the sol-gel process have been successfully used in drug delivery systems [18].

The aim of this chapter is to obtain a polymer system based on PVP, PVAl and chitosan containing pseudoboehmite nanoparticles, through the action of ionizing radiation, although the excellent biomedical properties in clinical practice of PVP-based hydrogels are confirmed, the difficult handling of these materials has been observed on their poor mechanical properties. Therefore, it is important to study new systems that maintain the properties required for these materials and at the same time to implement the physical:chemical and mechanical properties

2. Experimental study

2.1. Materials

The following reagents were used: aluminum nitrate $Al(NO_3)_3$, supplied by Dinâmica LTDA; ammonium hydroxide (NH_4OH), supplied by Audaz Reagente Tecnológico; polyvinyl alcohol, supplied by Bandeirante Química; poly(N-vinyl-2-pyrrolidone) (PVP), supplied by GAF Co.; chitosan, supplied by Polymar; poly(ethylene glycol), supplied by Oxiteno Brasil and agar supplied by Oxide.

2.2. Methods

The synthesis of nanoparticles of pseudoboehmite (PSB): The nanoparticles of the pseudoboehmite were obtained through the sol-gel process, according to Munhoz Jr. et al. [22], aluminum nitrate solution in H_2O, solution of ammonium nitrate in H_2O and solution of polyvinyl alcohol in H_2O, which is used to increase the viscosity of the aluminum nitrate solution. The solution of aluminum nitrate and polyvinyl alcohol was mixed with the ammonium hydroxide solution. The obtained product was washed with distilled water and dried through air.

Obtaining chitosan solution: The chitosan was dissolved in acetic acid solution, 2 wt% in H_2O, which was stirred for 48 h. After dissolution of chitosan, it was neutralized with 1 wt% sodium hydroxide solution in H_2O until pH 7.

Preparation of PVP/PVAl/PSB/chitosan membranes reinforced with PSB: Membranes with 3 wt% PSB were obtained. **Table 1** shows the composition of the membranes obtained.

The hydrogels were produced from a solution of PVP, PVAl, chitosan, agar and PEG in H_2O, which was heated and subsequently PSB was added to the solution. The solution was poured into polyethylene molds and after cooling, a physical gel was obtained.

Materials	Compositions (wt%)							
	Comp. 1	Comp. 2	Comp. 3	Comp. 4	Comp. 5	Comp. 6	Comp. 7	Comp. 8
PSB	0	0	0	0	0	0	3	3
PVP	10	10	2.5	2.5	2.5	2.5	2.5	2.5
PVAl	0	0	7.5	7.5	7.5	7.5	7.5	7.5
Chitosan	0	0	0	0	1	1	1	1
PEG	3	3	3	3	3	3	3	3
Agar	1	3	1	3	1	3	1	3
Water	86	84	86	84	85	83	82	80

Table 1. Membranes composition.

The physical gel was irradiated in Dynamitron electron beam, with energy of 1.5 MeV, with a dose of 25 kGy and a dose rate of 11.3 kGy/s, which promoted the formation of the cross-links and also the sterilization of the material.

2.2.1. Characterization of hydrogels

Hydrogels were characterized by visual test, mechanical test, gel fraction, thermal properties and swelling.

- *Mechanical test*: Tensile strength tests for the hydrated membrane were performed in a dynamometer of the Q-Test, model 65 J at 25 mm/min (ABNT-NBR 6241/80, with specimen type I).
- *Sol-gel fraction*: The samples were washed in Soxhlet extractors with boiling water for 36 h. The obtained gels were dried until reaching constant weight. The gel fraction was determined in relation to the initial weight of the sample according to Eq. (1).

$$\text{Sol} - \text{gel fraction (wt\%)} = (\text{Wfg}/\text{Wi}) \times 100 \qquad (1)$$

where Wfg = final weight (after drying); and Wi = initial weight of the sample.

- *Swelling*: The samples were maintained in water for 240 h. The water absorption was checked every hour step in the first 24 h. After that, each measurement was performed using 24-h steps until reaching constant weight. The hydration grade was determined by the difference of the weight before and after swelling according to Eq. (2).

$$\text{Swelling (wt\%)} = [(\text{Wfs} - \text{Wi})/\text{Wi}] \times 100 \qquad (2)$$

where Wfs = final weight (after swelling); and Wi = initial weight of the sample.

- *Thermal properties (DTA, TG and DTMA)*: Thermal analyses were performed by the Netzsch Thermische Analyze STA 409 equipment. The rate used in DTA and TG analysis was 10°C/min from ambient temperature until 600°C with 40 mL/min nitrogen flow. The thermodynamic-mechanical properties were determined in a Perkin Elmer equipment in the range from −80 to 0°C. The rate used in analysis was 1°C/min with 40 mL/min nitrogen flow.

3. Results and discussion

Visual characterization: **Figure 1** shows the obtained membranes with 25 kGy dose.

The membranes with the compositions 1, 2, 3 and 4 (**Figure 1**) are transparent, while others are translucent and slightly yellowish. Therefore, the presence of chitosan makes the hydrogels less transparent.

In the membranes with the compositions 1, 2, 3 and 4 were firm and without bubbles, the membranes of compositions 5, 6, 7 and 8 showed a greater adhesion or tack due to the presence of chitosan (**Figure 1**).

Tensile strength: **Table 2** and **Figures 2** and **3** present the results to the tensile strength tests to 7 days after irradiation.

The results show that comparing the obtained hydrogels, the Comp. 4 (based on PVP/PVAl/3wt% agar) and Comp. 3 (based on PVP/PVAl/chitosan/3wt% agar) (**Figure 2**) showed higher tensile strength and higher elongation.

The hydrogels Comp. 1 (PVP/1wt% agar), Comp. 2 (PVP/3wt% agar), Comp. 7 (PVP/PVAl/chitosan/PBS/1wt% agar) and Comp. 8 (PVP/PVAl/chitosan/PSB)/3wt% agar) exhibit lower tensile strength (**Figure 2**).

The Comp. 2 hydrogels (PVP/3 wt% agar) (**Figure 3**) show the smallest elongation.

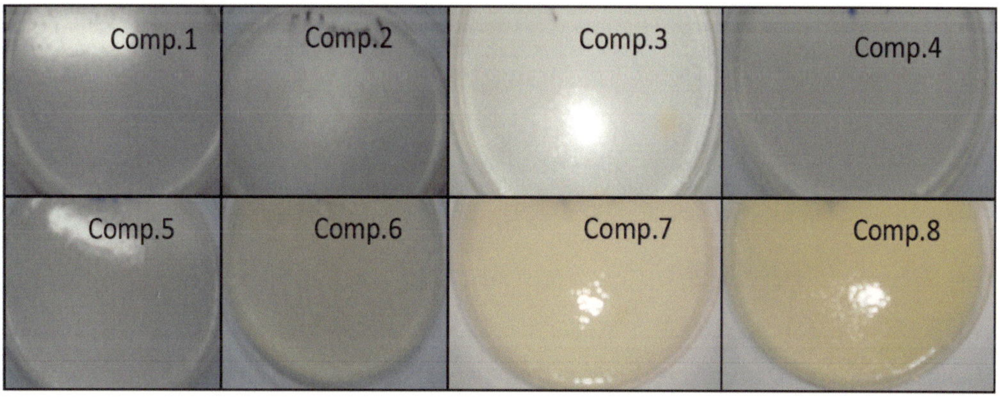

Figure 1. Comp. 1 (PVP/1wt% agar); Comp. 2 (PVP/3 wt% agar); Comp. 3 (PVP/PVAl/1 wt% agar); Comp. 4 (PVP/PVAl/3wt% agar), Comp. 5 (PVP/PVAl/chitosan/1 wt% agar), Comp. 6 (PVP/PVAl/chitosan/3wt% agar); Comp. 7 (PVP/PVAl/chitosan/PSB)/1wt% agar) and Comp. 8 (PVP/PVAl/chitosan/PSB/3 wt% agar).

Hydrogels	Tensile strength (MPa)	Elongation at rupture (%)
Comp. 1 (PVP/1wt% agar)	0.016 ± 0.006	136.±29
Comp. 2 (PVP/3wt% agar)	0.022 ± 0.006	42 ± 8
Comp. 3 (PVP/PVAl/1wt% agar)	0.108 ± 0.007	490 ± 70
Comp. 4 (PVP/PVAl/3wt% agar)	0.232 ± 0.020	568 ± 83
Comp. 5 (PVP/PVAl/chitosan/1wt% agar)	0.054 ± 0.008	289 ± 55
Comp. 6 (PVP/PVAl/chitosan/3wt% agar)	0.046 ± 0.005	496 ± 19
Comp. 7 (PVP/PVAl/chitosan/PSB/1wt% agar)	0.012 ± 0.007	227 ± 22
Comp. 8 (PVP/PVAl/chitosan/PSB/3wt% agar)	0.027 ± 0.007	339 ± 24

*The values obtained are the average of 16 experiments.

Table 2. Results of the tensile strength tests.

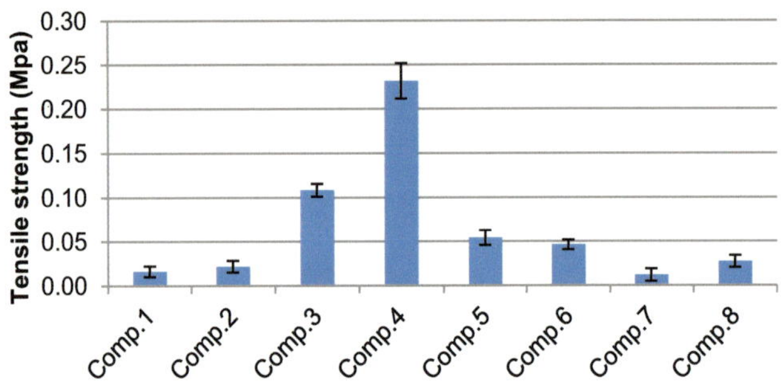

Figure 2. Results of the tensile strength.

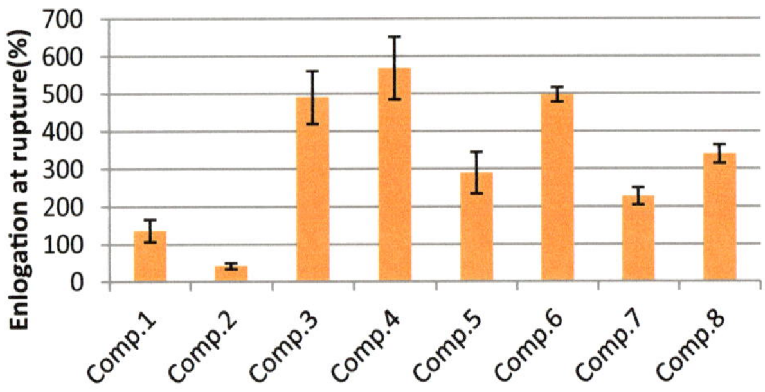

Figure 3. Results of elongation at rupture (%).

In general, hydrogels containing 3 wt% of agar in the composition, presented higher tensile strength and elongation at break, than their versions containing 1 wt% of agar (**Table 2**; **Figures 2** and **3**).

The data show that the association of chitosan with the increasing percentage of agar may have favored the elongation of the hydrogels (**Figure 3**);

The hydrogels containing pseudoboehmite presented lower results of tensile strength and lower results for elongation at rupture (**Table 2**; **Figures 2** and **3**).

Probably, the amount of pseudoboehmite present absorbs part of the free radicals during the irradiation process by decreasing the formation of cross-links.

Sol-gel fraction: **Table 3** and the **Figure 4** present the results obtained for the sol-gel fraction.

By results obtained in **Table 3** and **Figure 4**, it is observed that the hydrogels obtained with compositions 5, 6 and 7, containing PVAl, chitosan and pseudoboehmite, presented the smallest percentage of gel fraction. The presence of PVAl, chitosan and pseudoboehmite, probably, absorb part of the free radicals during the irradiation process, decreasing the formation of cross-links.

The hydrogels obtained with compositions 1 and 3 presented the highest percentages of gel fraction, although the conventional composition containing PVP/agar had a higher percentage of gel fraction than those containing PVP/agar/PVAl (**Figure 4**).

Hydrogels	Sol fraction (%)	Gel fraction (%)
Comp. 1 (PVP /1wt% agar)	15.37	84.63
Comp. 2 (PVP/3wt% agar)	43.27	56.73
Comp. 3 (PVP/PVAl/1wt% agar)	30.99	69.01
Comp. 4 (PVP/PVAl/3wt% agar)	44.38	55.52
Comp. 5 (PVP/PVAl/chitosan/1wt% agar)	54.05	45.95
Comp. 6 (PVP/PVAl/chitosan/3wt% agar)	61.85	38.15
Comp. 7 (PVP/PVAl/chitosan/PSB/1wt% agar)	57.53	42.47
Comp. 8 (PVP/PVAl/chitosan/PSB/3wt% agar)	44.63	55.37

*The values obtained are the average of 16 experiments.

Table 3. Results obtained for sol-gel fraction.

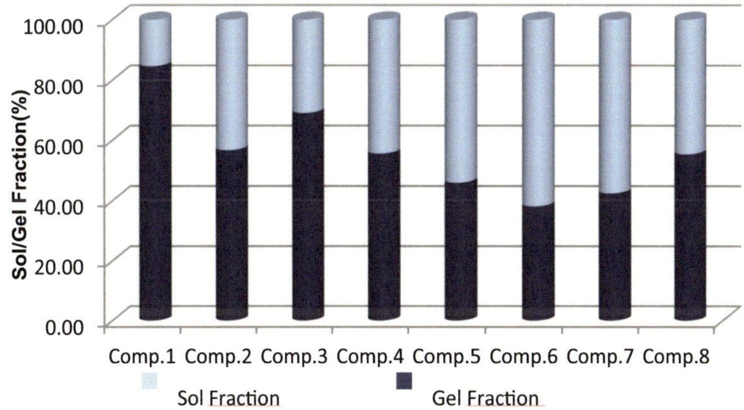

Figure 4. Results obtained for sol-gel fraction.

Comparing the hydrogels containing PVP/agar/PVAl with the hydrogels containing PVP/agar/PVAl/chitosan and PVP/agar/PVAl/chitosan/PSB, it is observed that the first have higher gel fraction indicating that the PVAl, probably absorbs a smaller part of the free radicals during the irradiation process than the chitosan and the pseudoboehmite (**Figure 4**).

The hydrogels obtained with compositions 2, 4 and 8 presented intermediate percentages of gel fraction. When comparing the agar concentration in the samples, it is observed that the increase in the agar concentration increases the gel fraction. Probably, the agar promotes the increase of the cross-link formation (**Figure 4**).

Comparing the results obtained for the gel fraction of the hydrogels containing chitosan and pseudoboehmite with those containing only one of these components, the values obtained were intermediates, indicating that these compounds acted independently of one another (**Figure 4**).

The Comp. 6 presents higher gel fraction. This sample does not contain pseudoboehmite in the composition (**Figure 4**).

Swelling: Swelling tests were performed for 7 and 30 days after irradiation of the hydrogels. **Figures 5** and **6** show the results obtained for the swelling tests.

The results show that hydrogels presented higher percentages of swelling after 30 days of irradiation (**Figures 5** and **6**).

The hydrogels based on only PVP and agar (Comp. 1 and Comp. 2) had a low swelling percentage after 7 days of irradiation (**Figure 5**), and after 30 days of irradiation had a significant increase in swelling percentage (**Figure 6**). Probably, some macroradicals are recombined during this stage.

For the PVP/PVAl/1 wt% Agar (Comp. 3) hydrogels, the swelling percentage was low after 7 days of irradiation and remained low after 30 days of irradiation (**Figures 5** and **6**). While for

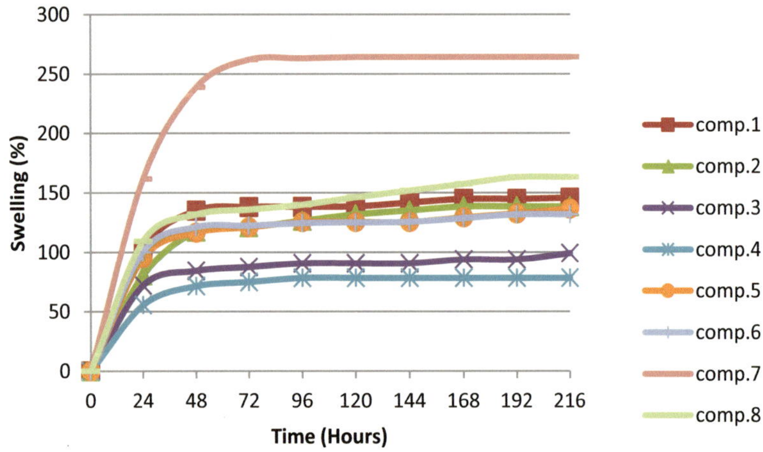

Figure 5. Results obtained for swelling tests of the hydrogels obtained after 7 days of the irradiation.

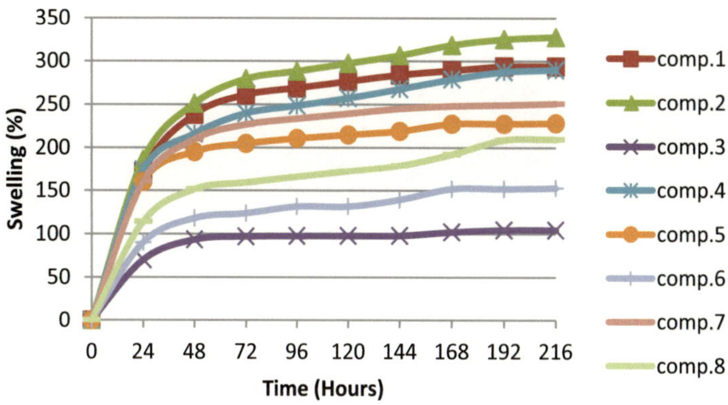

Figure 6. Results obtained for swelling tests of the hydrogels obtained after 30 days of the irradiation.

the hydrogels of similar composition only with variation in composition of 3 wt% agar, the increase in swelling percentage was significant. Probably, this difference is due to the presence of agar in a higher concentration that causes a greater absorption of water in the hydrogel structure because probably the agar promotes the increase of the cross-link formation.

The hydrogels containing PVP and PVAl, chitosan, Comp. 5 and Comp. 6 had low percentages of swelling after 7 days of irradiation (**Figure 5**), and the increase was not representative for the tests after 30 days of irradiation (**Figure 6**). Probably, the chitosan avoid the posterior cross-link.

The hydrogels containing PVP/PVAl/chitosan/PSB/1 wt% agar, Comp. 7, presented the highest percentage of swelling after 7 days of irradiation (**Figure 5**), but the increase was not significant after 30 days of irradiation (**Figure 6**). The hydrogels containing PVP/PVAl/chitosan/PBS/3 wt % agar, Comp. 8, presented low water absorption after 7 days of irradiation and significant increase after 30 days of being irradiated (**Figures 5** and **6**).

DTA and TG: The DTA and TG results for the obtained hydrogels are shown in **Figures 7–11**.

The results show that for the Comp. 1 hydrogels, containing PVP as the matrix, it can be observed that the PVP melts at a lower temperature, decreases the Tm and its degradation. For this composition, the Tg did not show considerable variation in relation to the pure PVP, and also its degradation temperature (**Figure 7**).

For the Comp. 2 hydrogels, where the matrix is also PVP, containing a higher percentage of agar, there was a decrease in Tg. PVP molecules have probably gained greater mobility at a lower temperature. However, T_m occurred at a higher temperature, indicating that degradation in the PVP molecules present in the hydrogel probably occurred. An increase occurred in temperature of degradation was observed (**Figure 7**).

For the Comp. 3 hydrogels, where the PVAl is found in greater proportion, the molecules gained greater mobility at lower temperature, that is, it decreased the Tg and the fusion occurred at lower temperature, and also, it decreased the Tm (**Figure 8**). The same occurred for the Comp. 4 hydrogels with PVP and PVAl matrix and 3 wt% agar (**Figure 8**), that compared with the other samples, the molecules gained greater mobility at lower temperature,

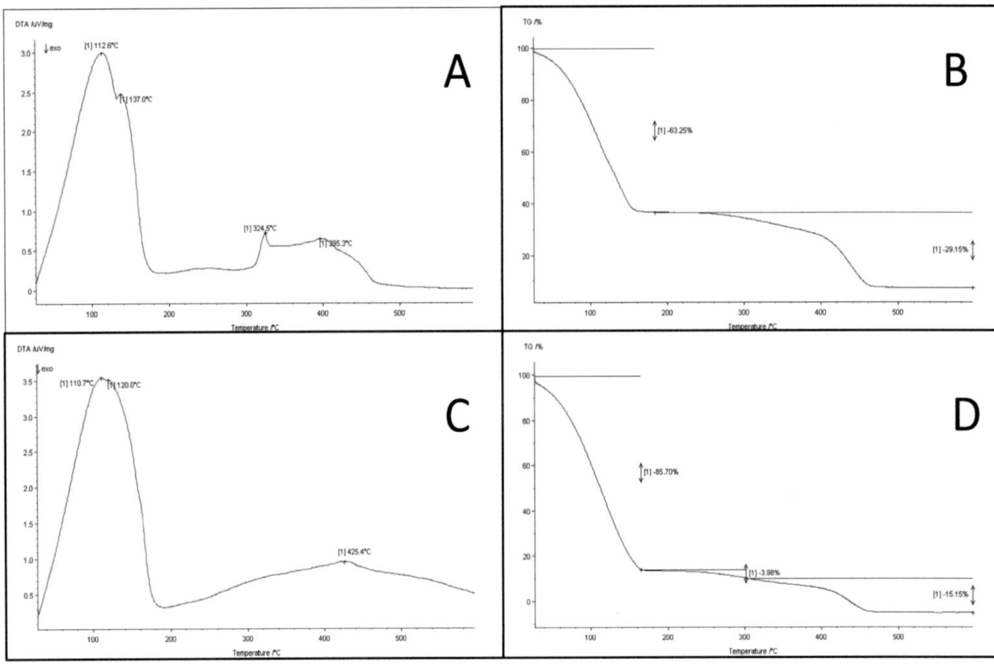

Figure 7. (A) DTA of pure PVP with 1 wt% of agar, (B) TG of pure PVP with 1 wt% of agar, (C) DTA of pure PVP with 3 wt% of agar; (D) TG of pure PVP with 3 wt% of agar.

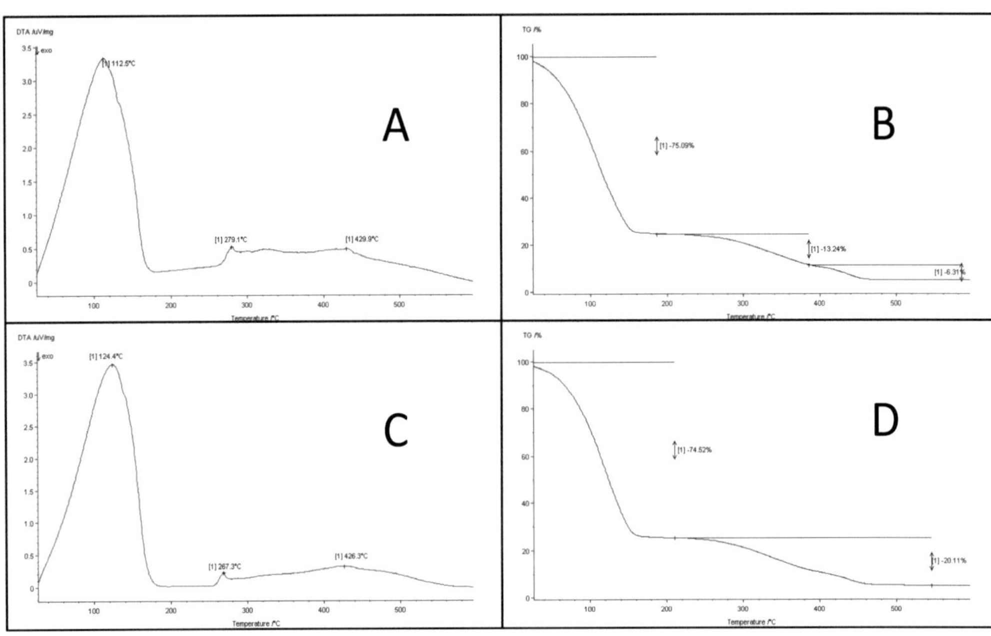

Figure 8. (A) DTA of the PVP/PVAl with 1 wt% of agar, (B) TG of the PVP/PVAl with 1 wt% of agar, (C) DTA of the PVP/PVAl with 3 wt% of agar and (D) TG of the PVP/PVAl with 3 wt% of agar.

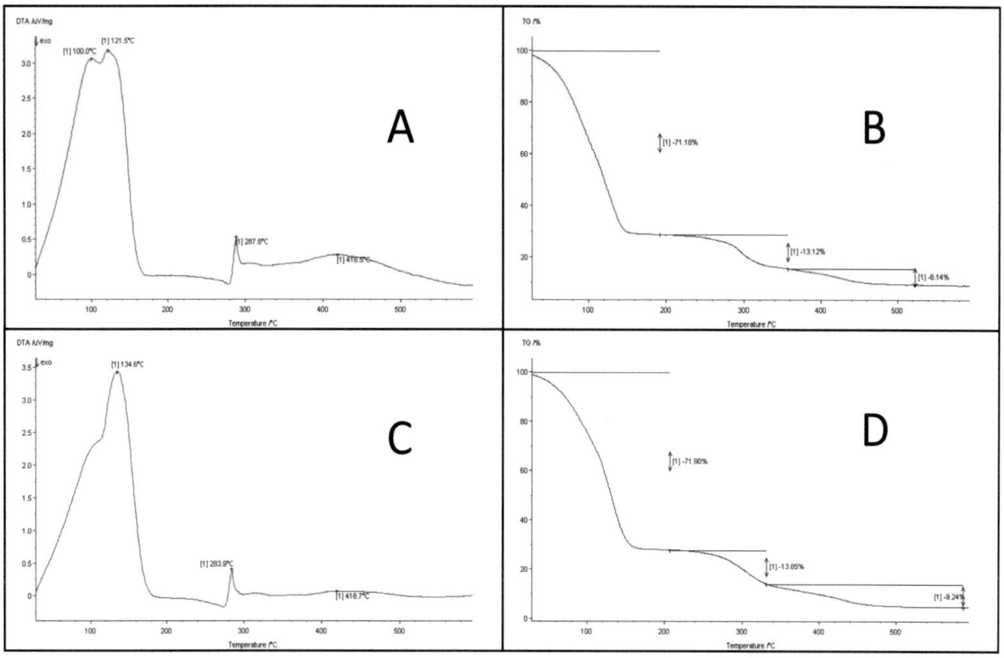

Figure 9. (A) DTA of the PVP/PVAl/chitosan with 1 wt% of agar, (B) TG of the PVP/PVAl/chitosan with 1 wt% of agar, (C) DTA of the PVP/PVAl/chitosan with 3 wt% of agar and (D) TG of the PVP/PVAl/chitosan with 3 wt% of agar.

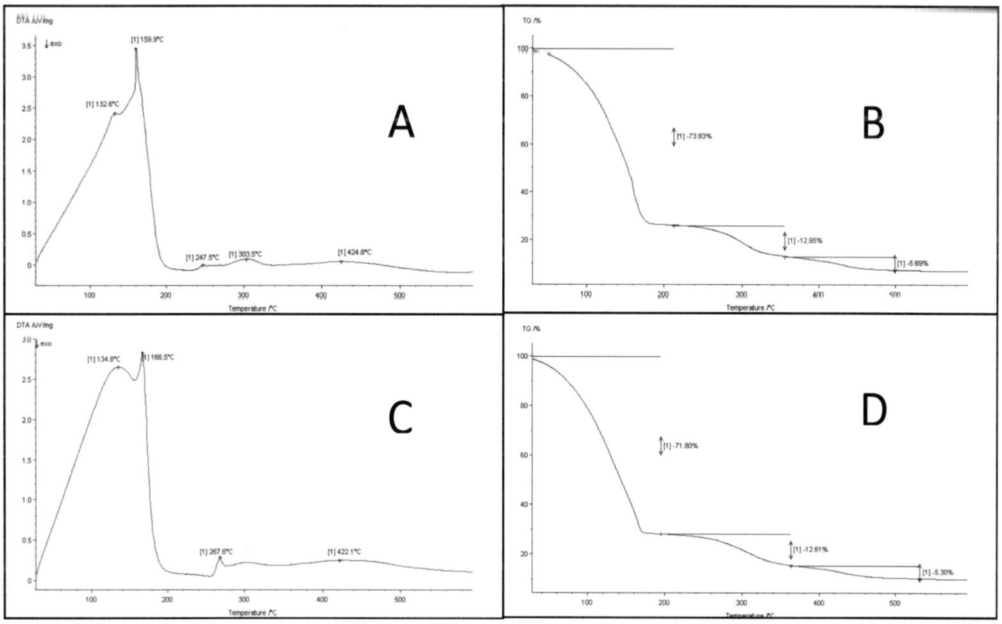

Figure 10. (A) DTA of the PVP/PVAl/chitosan/PSB with 1 wt% of agar; (B) TG of the PVP/PVAl/chitosan/PSB with 1 wt% of agar; (C) DTA of the PVP/PVAl/chitosan/PSB with 3 wt% of agar; and (D) TG of the PVP/PVAl/chitosan/PSB with 3 wt% of agar.

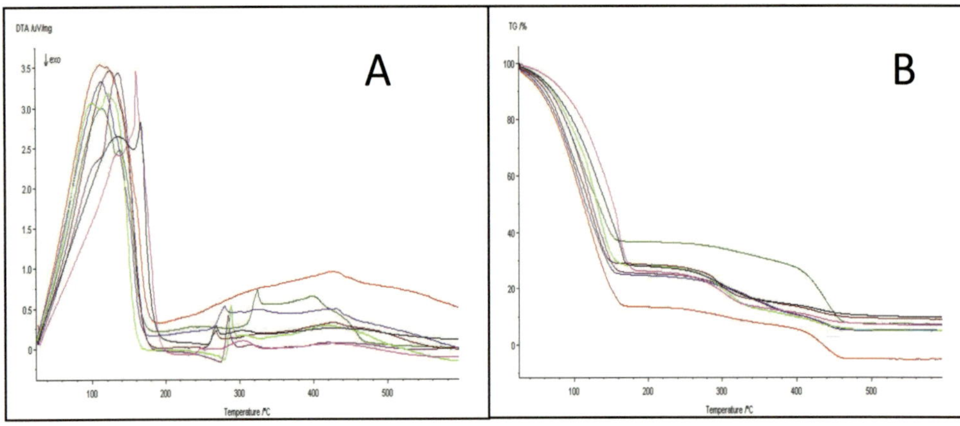

Figure 11. (A) DTAs and (B) TGs of the all compositions.

decreasing Tg and Tm at a lower temperature. It was observed that the decomposition occurred at a higher temperature for samples Comp. 3 and Comp. 4.

For membranes with composition 5 containing PVP/PVAl/chitosan/1 wt% agar, it was concluded that the water was not strongly retained and the molecules gained mobility at a lower temperature, decreasing Tg and Tm (**Figure 9**). However, the degradation was delayed, that is, it occurred at higher temperature than pure PVP (**Figure 7(A)** and **(B)**).

For Comp. 6 hydrogels with PVP/PVAl/chitosan/3 wt% agar, it can be concluded that the addition of agar did not hinder the loss of water. However, the molecules gained mobility at higher temperature (Tg higher), and the melting temperature was almost the same as temperature of sample 5. However, the degradation was delayed, occurring at a higher temperature (**Figure 9**).

DMTA: The DMTA results for the obtained hydrogels are shown in **Figures 12–15** and **Table 4**.

The results show that when comparing Comp. 1 and Comp. 2 hydrogels, the presence of a higher concentration of agar decreases the Tm value, causing an increase in the viscoelasticity of the material. This result can also be observed for the Comp. 4 hydrogels.

When comparing Comp. 1 and Comp. 3 hydrogels (**Figures 12** and **13**), the presence of PVAl causes an increase in T_m, and consequently, a decrease in the viscoelasticity of the material.

The presence of chitosan and pseudoboehmite in the hydrogels causes an increase in the T_m of the material, reducing its viscoelasticity (**Figures 14** and **15**). It is observed that the effect of chitosan is more effective in process. Probably, the chitosan structure, containing several hydroxyl groups (which may form hydrogen bonds), causes a bigger decrease in the viscoelasticity in the hydrogel of than the pseudoboehmite.

http://dx.doi.org/10.5772/intechopen.72007

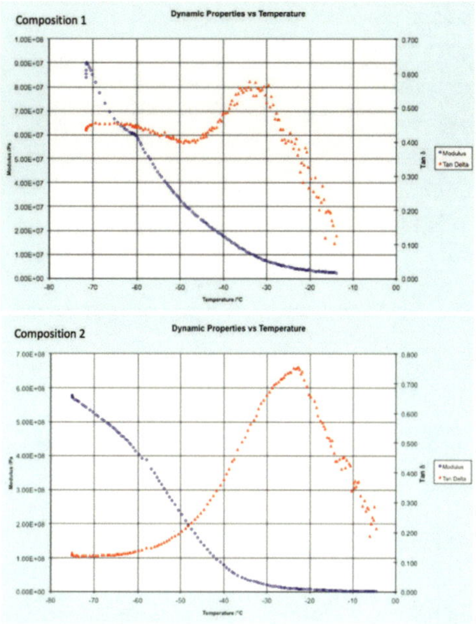

Figure 12. DMTA results of the Comp. 1 and Comp. 2.

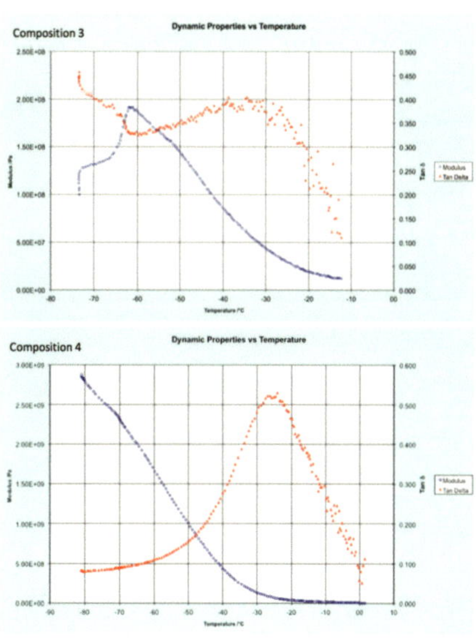

Figure 13. DMTA results of the Comp. 3 and Comp. 4.

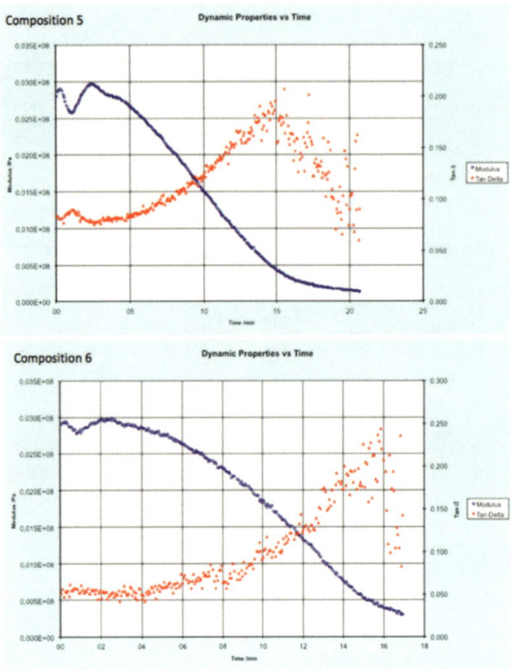

Figure 14. DMTA results of the Comp. 5 and Comp. 6.

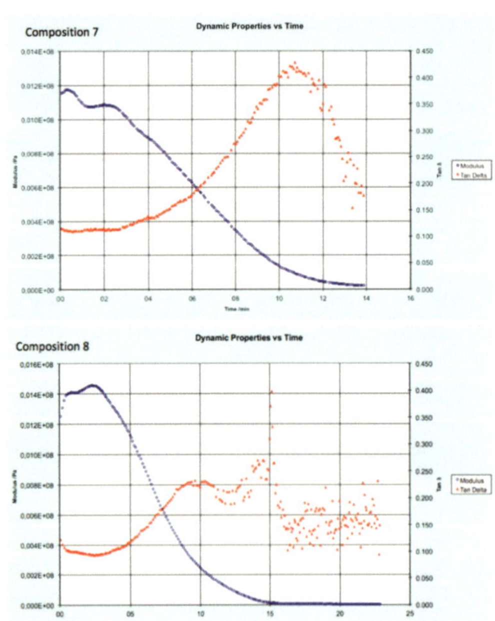

Figure 15. DMTA results of the Comp. 7 and Comp. 8.

Hydrogel	T_m (°C)	tan δ
Comp. 1 (PVP/1 wt% agar)	−33.9	0.5752
Comp. 2 (PVP/3 wt% agar)	−22.8	0.7540
Comp. 3 (PVP/PVAl/1 wt% agar)	−38.7	0.4045
Comp. 4 (PVP/PVAl/3 wt% agar)	−24.6	0.5202
Comp. 5 (PVP/PVAl/chitosan/1 wt% agar)	1.5	0.1953
Comp. 6 (PVP/PVAl/chitosan/3 wt% agar)	6.6	0.2439
Comp. 7 (PVP/PVAl/chitosan/PSB/1 wt% agar)	−24.5	0.4280
Comp. 8 (PVP/PVAl/chitosan/PSB/3 wt% agar)	2.7	0.2699

Table 4. Results obtained from T_m and tan δ of the studied hydrogels.

4. Conclusion

According to the results, the conclusions are as follows

- It is possible to obtain hydrogels based on PVP, PVAl and chitosan containing pseudoboehmite nanoparticles in the studied concentrations;
- The presence of chitosan makes the hydrogels less transparent and makes the hydrogels more adherent;
- The association of chitosan with increasing percentage of agar promotes the higher elongation at rupture of hydrogels;
- The hydrogels containing pseudoboehmite presented lower results of tensile strength and lower results for elongation at rupture. Probably, the pseudoboehmite absorbs part of the free radicals during the irradiation process by decreasing the formation of cross-links;
- The presence of PVAl increases the tensile strength and causes higher elongation at the rupture;
- In general, membranes containing 3 wt% of agar in the composition presented higher tensile strength and higher elongation at break than their versions containing 1 wt% of agar;
- The association of chitosan with increasing percentage of agar may have favored the elongation of the membranes;
- How much bigger the concentration of agar in the hydrogel, higher is the tensile strength and the higher is the elongation at break. Probably, the agar promotes the increase of the cross-link formation.
- The hydrogels containing PVAl, chitosan and pseudoboehmite presented the smallest percentage of gel fraction. The presence of PVAl, chitosan and pseudoboehmite, probably, absorb part of the free radicals during the irradiation process, decreasing the cross-links formation;
- All hydrogels' compositions studied showed higher percentages of swelling after 30 days of irradiation;

- Hydrogels based on PVP and agar had a low percentage of swelling after 7 days of irradiation, and after 30 days of irradiation, they had a significant increase in swelling percentage;
- For the hydrogel based on PVP, PVAl and agar, the presence of agar in a higher concentration promote a higher absorption of water in the hydrogel structure;
- The hydrogels containing PVP, PVAl, agar and chitosan had low percentages of swelling after 7 days of irradiation and the increase was not representative for the assay after 30 days of irradiation;
- The hydrogels based on PVP, PVAl, chitosan and pseudoboehmite containing agar in lower concentration presented the highest percentage of swelling after 7 days of irradiation, but the increase was not significant after 30 days of irradiation. The same hydrogels, containing agar in higher concentration presented low water absorption after 7 days of irradiation and significant increase after 30 days of irradiation;
- Comparing the hydrogels based on PVP and agar, the presence of a higher concentration of agar decreases the Tm, causing an increase in the viscoelasticity of the material. This result can also be observed for the hydrogels based on PVP, PVAl and agar;
- Comparing the hydrogels based on PVP and agar with the hydrogels based on PVP, PVAl and agar, the presence of PVAl causes an increase in Tm, that is, there is a decrease in the viscoelasticity of the material;
- The presence of chitosan and pseudoboehmite in the hydrogels causes an increase in the Tm of the material, reducing your viscoelasticity. It is observed that the effect of chitosan is greater. Probably, the structure of the chitosan, containing several hydroxyl groups promotes a pronounced decrease in the viscoelasticity of the material than the pseudoboehmite; and
- As the pseudoboehmite probably absorbs free radicals, to obtain hydrogels with higher mechanical properties, the dose irradiation must be increased to increase the density of cross-links.

Acknowledgements

This research has been supported by the Mack Pesquisa, Mackenzie Presbyterian University. We are grateful for the support.

Author details

Leila Figueiredo de Miranda*, Kátia Lucia Gonçalves Cunha, Isabella Tereza Ferro Barbosa, Terezinha Jocelen Masson and Antonio Hortêncio Munhoz Junior

*Address all correspondence to: leila.miranda@mackenzie.br

Mackenzie Presbyterian University, São Paulo, SP, Brazil

References

[1] Rosiak J, Rucinska-Rybus A, Pekala W. Polish Patent No. 151581 and also: US Patent No. 4871490; FRG Patent No. 3744289; GDR Patent No. 273200. Out.3 1989

[2] Miranda LF, Terence MC, Faldini SB. Processo de obtenção de membranas hidrofílicas a partir de poli(N-vinil-2-pirrolidona) PVP. 2009, Brasil Patente: Modelo de Utilidade. Número do registro: PI0900867-5, 16/04/2009

[3] Rosiak JM, Olejniczak J. Medical applications of radiation formed hydrogels. Radiation Physics and Chemistry. 1993;**42**(4-6):903-906

[4] Darwis D, Hilmy N, Hardningsih L, Erlinda T. Poly(N-vinylpyrrilidone) hydrogels: 1. Radiation polymerization and crosslinking of n-vinylpyrrolidone. Radiation Physics and Chemistry. 1993;**42**(4-6):907-910

[5] Bevington A. Comprehensive polymer science. In: Colin B, Colin P, editors. Interpretrating Polymer Network. 1990. pp. 423-436

[6] McCrum NG, Buckley CP, Bucknall CB. Principles of Polymers Engineering. Oxford: Oxford University Press; 1988

[7] Kamoun EA, Kenawy ES, Tamer TM, El-Meligy MA, Mohy Eldina MS. Poly (vinyl alcohol)-alginate physically crosslinked hydrogel membranes for wound dressing applications: Characterization and bio-evaluation. Arabian Journal of Chemistry. 2015;**8**(1):38–47

[8] Kamoun EA, Chen X, Mohy Eldin MS, Kenawy ES. Crosslinked poly(vinyl alcohol) hydrogels for wound dressing applications: A review of remarkably blended polymers. Arabian Journal of Chemistry. 2015;**8**(1):1-14

[9] Assis OB, Leoni AM. Filmes comestíveis de Quitosana. Revista Biotecnologia Ciência e Desenvolvimento. Edição nº30-Janeiro/Julho 2003. pp. 33-38

[10] Abbasipour M, Mirjalili M, Khajavi R, Majidi MM. Coated cotton gauze with Ag/ZnO/chitosan nanocomposite as a modern wound dressing. Journal of Engineered Fibers and Fabrics. 2014;**9**(1):124-130

[11] Gupta B, Agarwal R, Alam MS. Textile-based smart wound dressings. Indian Journal of Fibre & Textile Research. 2010;**35**:174-187

[12] Terence MC, Miranda LF, Faldini SB, Castro PJ. Efeito da adição de quitosana em blendas de poli(N-vinil-2-pirrolodona) e poli(álcool vinílico). In: 10° Congresso Brasileiro de Polímeros; 2009 Foz do Iguaçú; São Carlos: Associação Brasileira de Polímeros. 2009;**1**:1-7

[13] Ignatova M, Manolova N, Rashkov I. Novel antibacterial fibers of quaternized chitosan and poly(vinyl pyrrolidone) prepared by electrospinning. European Polymer Journal. 2007;**43**:1112-1122

[14] Dias AM, Braga ME, Seabra IJ, Ferreira P, Gil MH, De Sousa HC. Development of natural-based wound dressings impregnated with bioactive compounds and using supercritical carbon dioxide. International Journal of Pharmaceutics. 2011;**408**(1-2):9-19

[15] Moghadas B, Dashtimoghadam E, Mirzadeh H, Seidi F, Hasani-Sadrabadi MM. Novel chitosan-based nanobiohybrid membranes for wound dressing applications. RSC Advances. 2016;**6**:7701-7711

[16] Gonzalez JS, Maiolo AS, Ponce AG, Alvarez VA. Composites based on poly(vinyl alcohol) hydrogels for wound dressing. XVIII The Argentine Congress of Bioengineering and Clinical Engineering Conference VII, SABI 2011; 2011

[17] Yu H, Xu X, Chen X, Hao J, Jing X. Medicated wound dressings based on poly(vinyl alcohol)/poly(N-vinyl pyrrolidone)/chitosan hydrogels. Journal of Applied Polymer Science. 2006;**101**:2453-2463

[18] Munhoz Junior AH, Novickis RW, Miranda LF, Faldini SB, Terence MC, Ribeiro RR. Ceramic matrix for incorporating controlled release drugs, a tablet, method for obtaining the ceramic matrix and method for producing a tablet. US Patent 2011/0160230, Jun. 30, 2011. United States Patent and Trademark Office; 2011

[19] Kloprogge JT, Duong LV, Wood BJ, Frost RL. XPS study of the major minerals in bauxite, gibbsite and (pseudo)boehmite. Journal of Colloid and Interface Science. 2006;**296**: 572-576

[20] Munhoz AH, Meneghetti Peres R, Silveira LH, Andrade E Silva LG, Miranda LF. Irradiation of a nanocomposite of pseudoboehmite-nylon 6,12. Advances in Science and Technology. 2010;**71**:28-33

[21] AAS A, Kubota LT. A Utilização de materiais obtidos pelo processo de sol-gel na construção de biossensores. Química Nova. 2002;**25**(5):835-841

[22] Munhoz Jr. AH, Miranda LF, Uehara GN. Study of pseudoboehmite by sol-gel synthesis. AST-Advances in Science and Technology. 2006;**45**:260-265

Chapter 8

Development of PVA/Fe_3O_4 as Smart Magnetic Hydrogels for Biomedical Applications

Malik Anjelh Baqiya, Ahmad Taufiq, Sunaryono, Munaji, Dita Puspita Sari, Yanurita Dwihapsari and Darminto

Additional information is available at the end of the chapter

http://dx.doi.org/10.5772/intechopen.71964

Abstract

Polyvinyl alcohol (PVA)/Fe_3O_4 magnetic hydrogels had been fabricated by freezing-thawing (F-T) cycle technique, employing natural iron sand as the raw material for the magnetic micro- and nano-sized fillers. An exploration of the durability and magneto-elasticity as well as PVA hydrogel applications in the assessment of human brain tumor was also intensively conducted. The performance of the PVA and magnetic hydrogels mainly depends on the structural dynamic properties, such as polymeric crystallization and particle size. The durability of PVA/Fe_3O_4 magnetic hydrogels affecting the magnetoelasticity is determined by the concentration ratio of PVA and water, number of F-T cycles, and the concentration of Fe_3O_4 particles. By controlling those parameters, it was found that hydrogels had PVA: water ratio of 23:100 and four times F-T cycles possessed good mechanical properties. Due to the biocompatible character, the PVA hydrogel was used in the assessment of the human brain tumor, analyzed from the apparent diffusion coefficient (ADC) value representing the diffusion coefficient of a biological tissue. It was found that the abnormal tissue has a low ADC value compared with the normal one. Moreover, the higher b-value of the diffusion-weighted magnetic resonance imaging (DW-MRI) measurement is more preferred in obtaining a good contrast of the data imaging.

Keywords: PVA hydrogels, ferrogels, freezing-thawing method, magnetoelasticity, biomedical applications

1. Introduction

The hydrogel is one of the smart polymeric gels consisting of (physically or chemically) cross-linked polymer in water. Due to its hydrophobicity and biocompatibility properties, hydrogel has been an interesting biomaterial used commonly for biotechnological applications [1–3] and drug delivery systems [4–6]. The physically cross-linked hydrogels can be constructed by hydrogen bonds, crystallization, and ionic and protein interactions, whereas the chemically cross-linked hydrogels can be built by complex chemical reaction (aldehydes), high energy radiation, polymerization, and enzymes [1, 7–9]. There is a disadvantage for the hydrogels prepared by the chemically cross-linked agent and gamma irradiation, namely a toxic residue that might be harmful to biological tissue. Therefore, the physically cross-linked hydrogels are more preferable to be applied for biomedical purposes [8]. Basically, hydrogels are sensitive to some environmental variables, such as acidity level (pH), temperature, electromagnetic signal, light, pressure, and other stimuli, so that they can be applied based on the proper environmental condition [10, 11]. A review of a particular number of synthetic hydrogels for biomedical applications and tissue engineering has been discussed [4, 12, 13].

Polyvinyl alcohol (PVA) is a biocompatible, water-soluble, and nontoxic synthetic polymer, which can be prepared to be a flexible material called PVA hydrogel. A physically cross-linked PVA hydrogel can be achieved by freezing-thawing (F-T) cyclic process [14, 15]. The networking gel structure, crystallinity, stability, and viscoelastic properties of the PVA hydrogels have been investigated [16–19]. The properties of PVA hydrogels prepared by F-T process depend on the molecular weight and concentration of the aqueous PVA solution, temperature, time duration, and number of F-T cycle processes [20]. For instance, Li and coworkers have successfully produced a reversible gel using poly(N-isopropyl acrylamide) (PNIPA) and polyacrylamide (PAM), which can be controlled by external stimuli such as temperature [21].

It has been found that cross-linking density and crystallinity of PVA hydrogel influence the overall mechanical properties of a hydrogel. Gupta et al. [22] had studied the effect of PVA concentration on both modulus elasticity and tensile strength of PVA hydrogel. They found that both mechanical properties increased with increasing PVA concentration up to 16% due to a higher degree of crystallinity and developing hydrogen bond interaction in the PVA hydrogel. In contrast, it has also been shown that higher crystallinity of the hydrogel (obtained by increasing PVA concentration) may cause the increase of optical contact angle, indicating the decrease of water affinity [23]. This is one parameter that should be concerned for producing a stable PVA hydrogel. Moreover, it is revealed that the number of cyclic processes in the F-T method affects the degree of cross-linking density. A higher number of repeated cycle results in the decrease in the amount of not incorporated PVA in the networking structure of hydrogel meaning that the polymer chains of PVA are dispersed and unrelated each other [24].

In the tissue engineering, a transparent PVA hydrogel has been successfully developed as soft tissue substitution due to the similar microstructure and mechanical properties to that of the biological cells and organs [25]. PVA-based composite gels have been also intensively studied for wound healing, tissue replacement, and magnetic-controlled drug delivery devices [26–28]. Liu et al. [29] have demonstrated PVA hydrogel produced by the F-T process as one of the potential materials for an artificial blood vessel. Furthermore, the F-T process can be used widely for preparing and storing cell-laden hydrogels with adjustable mechanical properties [30].

One important key for organ replication in tissue engineering is a complete knowledge of microscopic, physical, and chemical properties based on the desired organ. Recent development of hydrogel technologies for mimicking natural tissue has been briefly reviewed [31]. It is crucial for constructing a tissue from hydrogel without causing significant cell damage. Investigation on a transparent PVA hydrogel as tissue-equivalent material in the surgical application has been conducted [32]. They have found that PVA hydrogels could be applied for a suitable substitution for soft tissue accuracy and surgical purposes. Forte et al. [33] have successfully developed a composite hydrogel (PVA hydrogel) to mimic brain tissue. They tried to vary the PVA concentration in order to tune the mechanical response of brain tissue within a wide range of stress/strain and testing conditions. In the brain tissue replica, it is important to study accurately the brain shift phenomenon between the abnormal and the normal one.

2. Magnetic hydrogels (Ferrogels) and their physical properties

The magnetic hydrogel, or so-called ferrogel, is one of the "smart" polymeric composite gels containing micro- or nano-sized magnetic particles as filler in its polymeric matrix. There are some magnetite hydrogels (ferrogels) that have been successfully developed recently, namely PVA-Fe_3O_4-based hydrogel [34], Fe_3O_4-polyacrylamide (PAM) hydrogels [35], and Fe_3O_4-polymethylmethacrylate (PMMA) hydrogels [36]. They have found that those magnetic nanocomposites forming magnetic hydrogels have superparamagnetic behavior due to the presence of dispersed magnetic nanoparticles in the polymeric matrix. The properties of superparamagnetic of magnetic iron oxide nanoparticles itself have been investigated intensively [37]. A simulation study of deformation, elasticity, and magnetic response of magnetic nanoparticles cross-linked in a gel (polymeric matrix) has been conducted [38]. They have found that the degree of networking chains plays an important role in determining the stiffness and magnetosensitivity of the magnetic hydrogels. The sensitivity of magnetic response in the external magnetic field depends strongly on the volume fraction of both magnetic nanoparticles and polymeric base matrix influencing the interacting energy in the ferrogel [28]. They found an optimum magnetosensitivity with Fe_3O_4 and PVA concentration in the range of 17–34% and 10–12.5%, respectively.

Ferrogel is a new type of polymeric matrix composites, in which they are physically (or chemically) cross-linked polymer network containing dispersed magnetic particles. Zrínyi et al. [39] have synthesized ferrogel as a new promising material for magnetic-responsive applications. Ramanujan and Lao [40] and Reséndiz-Hernández et al. [41] have developed composite gels based on PVA and magnetite (Fe_3O_4) particles by conventional F-T process. Moreover, Hernández et al. [19] have reported the viscoelastic properties of PVA hydrogel and ferrogel prepared by F-T cyclic process. It has been revealed that the reinforcement effect from the magnetic particles, as well as the mechanical properties, of the PVA-Fe_3O_4 ferrogels depends on the size, possible agglomeration, and concentration (volume fraction) of the magnetic particles [42]. It has been noted that the concentration of Fe_3O_4 nanoparticles affects the thermal stability of the ferrogels [43]. As for the mechanical properties of the PVA-Fe_3O_4 ferrogels, it has been shown that the deflection and elongation parameters are dependent on the Fe_3O_4 concentration and external magnetic field strength [40] and the deformation is independent on the shape of ferrogels [44]. Experimentally, the magnetodeformation of the ferrogels with highly concentrated particles (approximately above 30%) is due to the effect of short range order and magnetic interaction among the particles [45, 46]. Furthermore, the structural and magnetic behavior of ferrogels has been intensively studied with the dispersed magnetic particle having average size less than 10 nm [47, 48].

Sunaryono et al. have shown that the hydrogels owing the threshold PVA concentration of 23% in water content have the best mechanical properties [49]. Moreover, the lower Fe_3O_4 concentration has been found to be responsible for a weak magnetic response due to the increase of particle free volume and the decrease of interaction energy between magnetic nanoparticles and the cross-linked PVA hydrogel [48]. However, Sunaryono et al. [48, 49] stated a crucial problem is related to the durability of the PVA hydrogels and ferrogels prepared by F-T cyclic process.

In this chapter, at first, we provide a study of PVA hydrogel application in tissue engineering. Then, it is continued by investigation of the durability of ferrogels prepared by F-T cyclic process. Finally, the structural and magnetic properties of ferrogels are discussed briefly.

3. Fabrication and characterization of PVA hydrogel

The PVA polymeric powder (Mw = 60,000 g/mol, Merck Schuchardt OHG, Germany) with a degree of hydrolysis ≥98% was used for PVA solution. PVA hydrogels were fabricated by F-T cyclic process. First, the PVA polymer was dissolved in distilled water with a variation of weight compositions, namely 7.5, 10, 12.5, and 15 wt%. The solution was mixed and heated at 70–90°C to improve the solubility of the PVA polymer in water, as suggested in the previous papers [48, 50, 51]. Once the mixture was perfectly dissolved, indicated by a physical change from liquid to paste, it was then loaded into a cylindrical plastic mold followed by F-T process. The solution was cooled and kept at the frozen state at –10°C for 3 h. The process was continued by thawing at room temperature for 1 h. This F-T process was repeated to obtain PVA hydrogel samples up to five cycles. The PVA hydrogel samples were also prepared by varying the composition ratio of PVA and water as mentioned earlier.

The 1.5-T scanner (Signa Horizon; GE Medical Systems, Milwaukee, WI, USA) was used for the study of diffusion-weighted magnetic resonance imaging (DW-MRI). Apparent diffusion coefficient (ADC) value was obtained from MRI with diffusion-weighted imaging (DWI) method following Stejskal-Tanner sequence. The ADC value was calculated by Functool software (GE Medical Systems) for each sample. The characterization steps were similar to the previous reports [50, 51].

The consistency measurement was conducted using a penetrometer (Precision 73,515, Petroleum Analyzer Co., San Antonio, TX, USA) using a pressure sensor. The penetrometer was set under gravity force for 5 s, and the depth of penetration was measured in tenths of millimeters. The depth of penetration depended on the kinetic energy applied to the penetrometer and the sample resistance. The resistance data were collected to obtain the consistency value describing the required mechanical force to decelerate from its initial velocity to zero velocity.

4. Preparation of Fe_3O_4 nanoparticles from iron sand

Fe_3O_4 nanoparticles were prepared by coprecipitation method employing natural iron sand as a raw material. The preparation was the same as explained in the former papers [48, 52]. First, iron sand was extracted by permanent magnet several times to obtain microsized Fe_3O_4 powders. HCl and NH_4OH were used as dissolving and precipitating agents, respectively. Fe_3O_4 nanoparticles produced by the coprecipitation method were based on the following chemical reaction [52, 53].

$$Fe_3O_4 + 8HCl \rightarrow 2\,FeCl_3 + FeCl_2 + 4H_2O \tag{1}$$

$$2\,FeCl_3 + FeCl_2 + 8NH_4OH \rightarrow Fe_3O_4 + 8NH_4Cl + 4H_2O \tag{2}$$

Both reactions were maintained at room temperature. A complete reaction was indicated by the color change of the solution and the formation of black precipitation. Finally, the precipitated powders were rinsed several times using distilled water and then dried at 100°C for 1 h for ferrogel fabrication.

5. Fabrication and characterization of PVA/Fe_3O_4 hydrogel (Ferrogel) based on iron sand

Ferrogel was fabricated by distributing the prepared Fe_3O_4 nanoparticles in the PVA hydrogel paste solution, and then, they were stirred to obtain a uniform gel. Furthermore, the mixture gel was placed into a cylindrical mold to perform the similar F-T cyclic process as in the PVA hydrogels fabrication. The ferrogel samples were prepared by varying the concentration of PVA and Fe_3O_4 nanoparticles, as well as the number of F-T cycles.

Basic characterizations using X-ray diffractometer (XRD) and transmission electron microscopy (TEM) were conducted to analyze the crystal structure and particle morphology of Fe_3O_4 nanoparticles and ferrogels, respectively. Vibrating sample magnetometer (VSM) and superconducting quantum interference device (SQUID) measurements were taken to investigate the magnetic properties of the PVA ferrogels. Particle size and the distribution of Fe_3O_4 nanoparticles in the PVA hydrogels were analyzed using small-angle X-ray scattering (SAXS) instrument as in the reported paper [48].

The ferrogels were exposed to the external magnetic field of an electromagnet apparatus, which is able to generate a magnetic field up to 460 mT. The response of the ferrogel was measured by the extent of deflection and elongation. The top end of a ferrogel sample was fixed, whereas the lower end was free to deflect and elongate during the application of the external magnetic field. Variation in the magnetic field was obtained by changing the electric current of the electromagnetic apparatus. The Young's modulus was measured using a universal mechanical tester.

6. Applications of PVA hydrogels for tissue engineering

PVA hydrogels fabricated through a number of F-T cycle processes have the ability to mimic a complex structure of human body. In a normal condition, water can be diffused into organic tissue during body's metabolism. However, the diffusion of water can be disturbed if water molecules are passing through an abnormal tissue due to swelling of the tissue, for example in tumor tissue. Moreover, about two-thirds of the human body consist of water, in which water molecules have hydrogen atoms that allow for magnetic resonance imaging (MRI) observation. Diffusion-weighted MRI (DW-MRI) is a device with high sensitivity in detecting "Brownian motions." The diffusion of water molecules caused by heat energy associated with the temperature of the human body can be used for analyzing a variety of brain diseases including brain tumors [54]. Additionally, diffusion-weighted imaging (DWI) is a technique that can be used to measure diffusion of water molecules in biological tissue such as white matter in the brain. In the MRI observation, the value of apparent diffusion coefficient (ADC) is used widely for describing the diffusion coefficient of the material.

In order to investigate the diffusion properties and the consistencies of the fabricated PVA hydrogels applied for tissue replica, several PVA hydrogel samples have been produced by variation number of F-T cycles and PVA concentrations. ADC values of the PVA hydrogels were obtained by performing the DW-MRI measurement. Sari et al. [51] have shown the ADC values versus PVA concentration for PVA hydrogels with the F-T process of three to five cycles. It has been found that the increase of PVA concentration from 7.5 to 15 wt% decreases the ADC values corresponding to the diffusion coefficient of all PVA hydrogels.

The enlargement of the tumor cells causes a reduction in the volume of extracellular space, increases the intracellular viscosity, and then inhibits the movement of water molecules described by the decrease of ADC value. Moreover, for PVA hydrogel fabricated by cryogelation process, the decrease of ADC value is described by swelling indicated by the significant increase of crystallization of hydrogel with increasing the number of F-T cycles. ADC value helps to distinguish a tumor tissue from a nontumor tissue. However, the abnormal tissue of brain tumor has a variety of classifications depending on the location and type of tumor tissue. Therefore, in the application of DW-MRI, it is necessary to find the most aggressive area at first to identify the highest cellularity and the most restrictive for the movement of water molecules. The use of higher b-value is to obtain a brighter contrast and has the impact on the easiness of diffusion imaging. The higher b-value can produce images on the high value of signal-to-noise ratio (SNR). Otherwise, at 1.5 T or lower, a low b-value may produce a poor image quality and lower value of SNR [55].

Figure 1 displays the ADC values of the PVA hydrogels measured from DW-MRI with b = 1000 and 3000 s/mm^2 as a function of F-T cycles and different PVA concentration. It has been found that the greater the concentration and number of cycles, the lower the diffusion coefficient of the PVA hydrogels. The crystallization, the degree of the physically cross-linked network, and the stiffness of hydrogel increase with the increase of the F-T cycle [15, 56]. The increase of crystallization indicates that the diffusion of water may be inhibited, so that the value of diffusion coefficient, described by ADC value, decreases. This is general diffusion behavior of a hydrogel, in which the diffusivity of a hydrogel decreases as cross-linking density increases and as the volume fraction of water within the hydrogel decreases [57]. It has also been indicated that the linearity of the ADC value as a function of F-T cycles at b = 3000 s/mm^2 is better than that at b = 1000 s/mm^2. This result is in a good agreement with the former result [55].

Figure 2 shows the consistency measurement as a function of F-T cycles and different PVA concentration. The data show that the higher PVA concentration and a number of F-T cycles cause the lower consistency and ADC value [50]. These results are consistent with the former paper [58]. The Pearson correlation method was used to correlate the data and are presented in **Table 1**. It is shown that the average value of ADC at b = 3000 s/mm^2 is good and slightly smaller than that at b = 1000 s/mm^2. The data have a good correlation (correlation number of 0.92—0.99), so that it can assess the abnormal tissue consistency [51].

Generally, ADC values of the human brain for both normal and abnormal cases are different significantly. In the DW-MRI analysis, ADC value in the normal human brain is about 0.75 mm^2/s and the higher b-value results in the lower ADC value. A tissue having low ADC value eliminates signals faster than that on the tissue having higher ADC value, and therefore, the contrast should increase. Sari et al. [51] have analyzed some cases for human brain tumor from the DW-MRI images at b-value of 1000 and 3000 s/mm^2. They also found that the tissue having low ADC value indicates lower consistency or harder than the tissue having high ADC value. The ADC measurement using b = 1000 s/mm^2 can distinguish the harder tissue with the

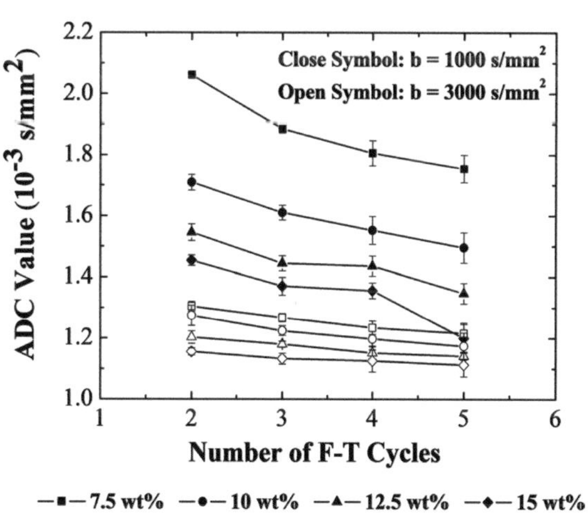

Figure 1. ADC value of PVA hydrogels on the DW-MRI at b-value of 1000 s/mm^2 (closed symbols) and 3000 s/mm^2 (opened symbols) as a function of the number of F-T cycles and PVA concentration.

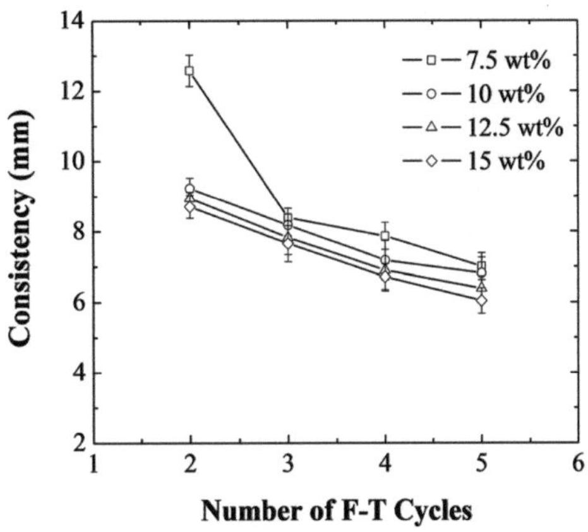

Figure 2. Consistency measurement using digital penetrometer as a function of the number of F-T cycles and PVA concentration.

normal one and provides a clearer image, although the ratio of normal and abnormal tissue is not as high as the use of b = 3000 s/mm². A better value of correlation with the physical parameters gives a suggestion that the use of DW-MRI 1.5 T with b = 1000 s/mm² provides a better image and the use of penetrometer is necessary for additional information for determining surgery. Otherwise, the use of DW-MRI 1.5 T with b-value higher than 1000 s/mm² is more preferred to examine the swelling that occurs around the area of the abnormal tissue because it provides more contrast image.

PVA hydrogels with different F-T cycle at constant PVA concentration of 10 wt%	**Data (s/mm²)**	
	b = 1000	**b = 3000**
Two cycles	0.96	0.77
Three cycles	0.96	0.99
Four cycles	0.99	0.96
Five cycles	0.96	0.97
PVA hydrogels with different PVA concentration at constant F-T cycle for three times	Data (s/mm²)	
	b = 1000	b = 3000
7.5 wt%	0.98	0.93
10 wt%	0.99	0.99
12.5 wt%	0.94	0.99
15 wt%	0.92	0.98

Table 1. The Pearson correlation result for both data of PVA hydrogels with different F-T cycle and PVA concentration.

7. Stability and durability of PVA-Fe_3O_4 hydrogels (Ferrogels)

In order to study the stability and durability of ferrogel, a number of ferrogel samples have been fabricated with a variation of PVA and water ratio, the concentration of Fe_3O_4 nanoparticles, and a number of F-T cycles. The stability was investigated by observing the increase of the required external magnetic field to elongate and deflect the ferrogels until a certain length and distance over a particular time. The observations were conducted from the first day since the ferrogels fabricated until the fifth day. The ferrogels can be called relatively stable if the change of required magnetic field is considerably small over the time to get the same deformation condition.

Table 2 shows ferrogel samples with a variation of PVA and water ratio together with their elasticity moduli. It can be seen from the elasticity properties that the stiffness of ferrogel depends on the PVA concentration in water. Higher PVA concentration causes the increase of stiffness. The stability of ferrogels is shown in **Figure 3. Figure 3** demonstrates the time dependence of the required external magnetic field to elongate ferrogel up to 1 mm. It indicates that

No.	Ratio of PVA and water	Fe_3O_4 (wt%)	Number of F-T cycles	Sample code	Modulus elasticity (Pa)
1	13:100	10	4	PA.13	67.18
2	18:100	10	4	PA.18	69.96
3	23:100	10	4	PA.23	61.23
4	28:100	10	4	PA.28	119.10
5	33:100	10	4	PA.33	221.20

Table 2. Ferrogel samples prepared with different PVA and water ratio, together with the modulus elasticity.

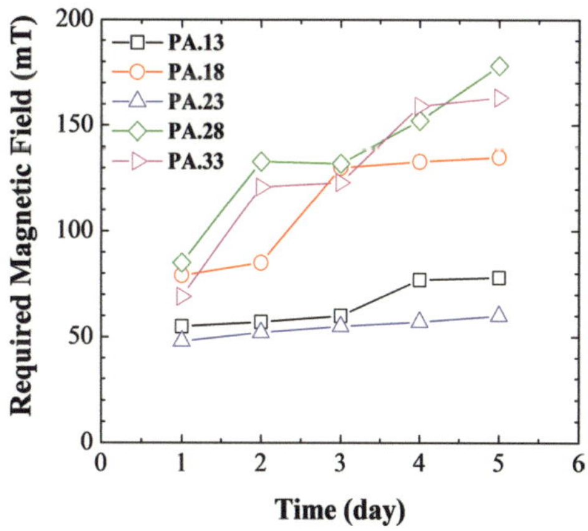

Figure 3. Required magnetic field to elongate ferrogel up to 1 mm as a function of time for ferrogels with different PVA and water ratio.

ferrogel becomes stiffer with the passage of time due to the decrease of water content. The ferrogel with PVA and water ratio of 23:100 (PA.23) shows a relatively small change of the required magnetic field up to five days indicating a relative stability compared to the others. This stability relates to the optimum portion of water inclusion binding in the PVA hydrogel.

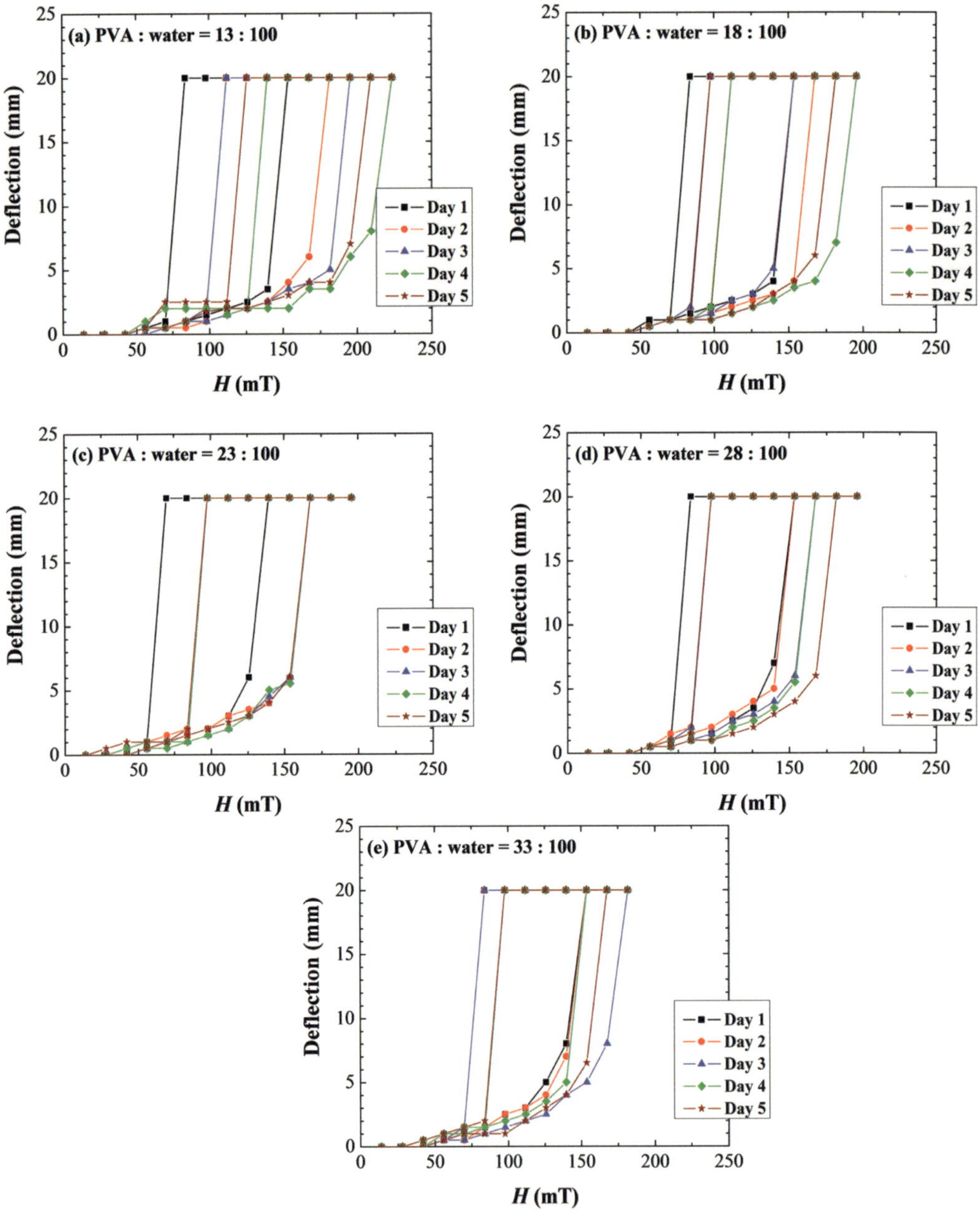

Figure 4. Hysteresis curves of the deflection behavior of ferrogels with PVA and water ratio: (a) 13:100, (b) 18:100, (c) 23:100, (d) 28:100, and (e) 33:100.

Figure 4 presents magnetic field dependence of ferrogel deflection with different PVA and water ratio. **Figure 4** displays interesting hysteresis loop in which the deflection increases with increasing magnetic field and returns to its original length through a different path, thereby decreasing the magnetic field. These noncontinuous deflection behaviors have also been observed by Zrínyi et al. [59] and modeled by Snyder et al. [60]. The hysteresis loops tend to shift day by day, indicating a rigid gel character due to the decrease of water content.

Ferrogels with different concentration of Fe_3O_4 nanoparticles and the modulus elasticity are presented in **Table 3**. It appears that there is no significant change in the modulus of elasticity with the increase of Fe_3O_4 concentration from 5 to 12.5 wt%. This result is consistent with the former report [40], in which the obtained elastic modulus was in the range of 0.17–0.75 MPa for a magnetoactive elastomer. **Figure 5** shows the stability characteristic of the ferrogels with various Fe_3O_4 concentrations associated with **Table 3**. It implies that the increase of nano-sized Fe_3O_4 concentration tends to decrease the required magnetic field to elongate ferrogel up to the same length for each day, indicating the decrease of water content and stiffer ferrogels. Ferrogel with Fe_3O_4 concentration of 10 wt% (FP.10) appears to be moderately stable compared to the others. For the ferrogels with Fe_3O_4 concentration less than 10 wt%, the trapped magnetic particles in the PVA chain were less and therefore the distribution was inhomogeneous, creating more spaces which were filled with water. According to the structural model of hydrogel described by Goiti et al. [43], the trapped, free and linked water molecules attached to the PVA chain may cause a soft ferrogel and dry quickly due to rapid evaporation of the water. On the other hand, for ferrogels with Fe_3O_4 concentration more than 10 wt%, there might be a space filled by Fe_3O_4 nanoparticles so that the water is suppressed. The Fe_3O_4 nanoparticles could directly coincide with the polymer chain. For the ferrogel with Fe_3O_4 concentration of 10 wt%, it is expected to have a proportional amount of solids and liquid, resulting in a good cross-linked hydrogel and then the trapped water can maintain flexibility of the gel.

Figure 6 shows magnetic field dependence of ferrogel deflection with different concentration of Fe_3O_4 nanoparticles. The hysteresis behavior of the deflection curves depends on the concentration of Fe_3O_4 nanoparticles and the elasticity of ferrogel itself. It should be noted that the hysteresis behavior observed in ferrogel is not consequences from the magnetic particles [59]. It appears that ferrogels with Fe_3O_4 concentrations of 10 and 12.5 wt% have no significant change in the hysteresis loops up to the fifth day. The change and the distortion

No	Ratio of PVA and water	Fe_3O_4 (wt%)	Number of F-T cycles	Sample code	Modulus elasticity (Pa)
1	23:100	5	4	FE.5	60.54
2	23:100	7.5	4	FE.7	60.94
3	23:100	10	4	FE.10	61.23
4	23:100	12.5	4	FE.12	61.59

Table 3. Ferrogel samples prepared with different Fe_3O_4 concentrations, together with the modulus elasticity.

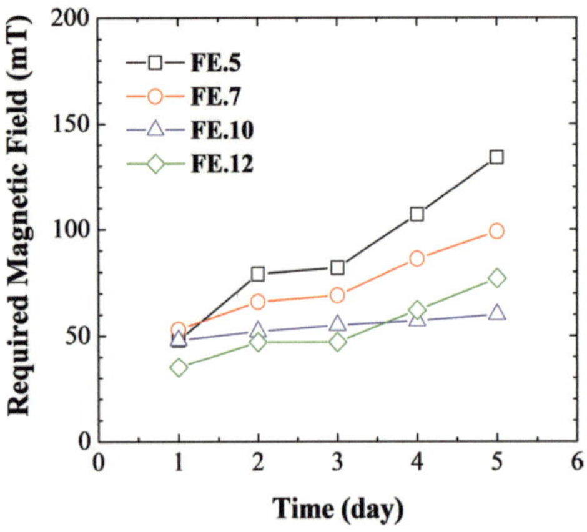

Figure 5. Required magnetic field to elongate ferrogel up to 1 mm as a function of time for ferrogels with different Fe_3O_4 concentrations.

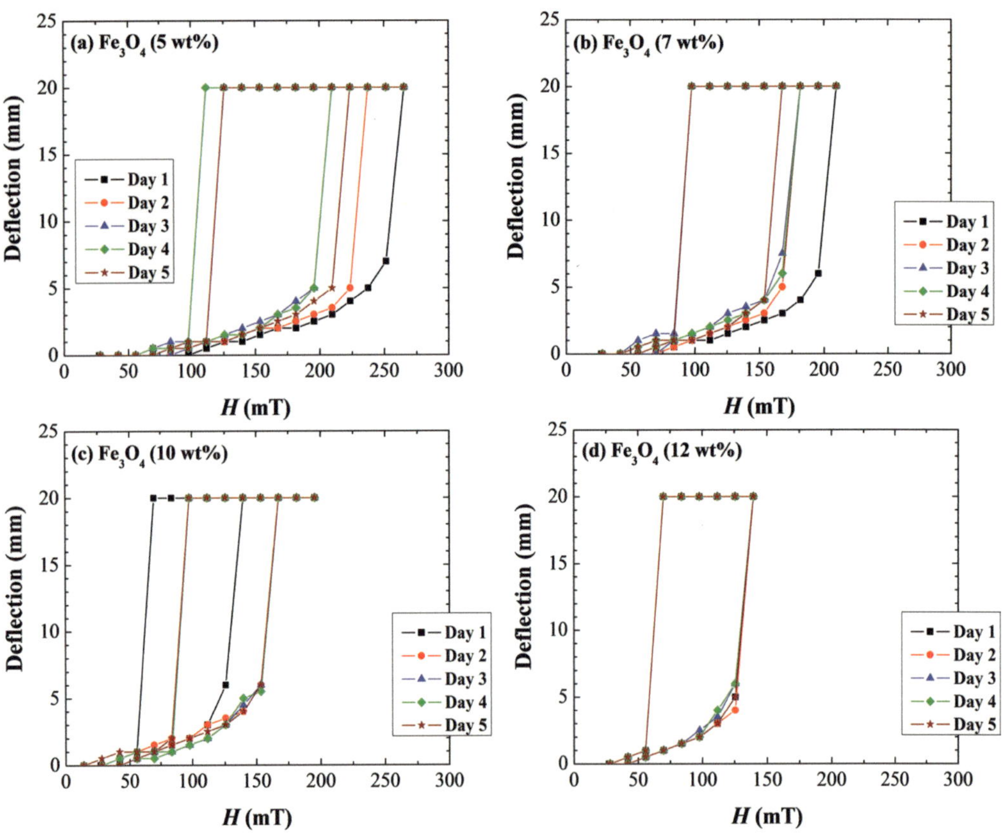

Figure 6. Hysteresis curves of the deflection behavior of ferrogels with Fe_3O_4 concentration of: (a) 5 wt%, (b) 7 wt%, (c) 10 wt%, and (d) 12 wt%.

No	Ratio of PVA and water	Fe_3O_4 (wt%)	Number of F-T cycles	Sample code	Modulus elasticity (Pa)
1	23:100	10	4	FT.4	61.23
2	23:100	10	8	FT.8	535.2
3	23:100	10	12	FT.12	267.0

Table 4. Ferrogel samples prepared with a different number of F-T cycles, together with the modulus elasticity.

of hysteresis shape are influenced by magnetostatic and magnetostriction mechanisms in the ferrogel [61]. Sample geometry may also be one parameter for determining the mechanical behaviors (elongation, deflection, etc.) of the ferrogel in the external magnetic field [60].

The modulus elasticity of ferrogels with a different number of F-T cycles was also investigated as shown in **Table 4**. In general, the greater the number of F-T cycles, the more rigid ferrogels will be obtained due to the evaporation of water. **Figure 7** shows the durability of ferrogels produced by 4, 8, and 12 times F-T cycles. It is also clear that ferrogel produced by four times F-T cycles has better stability as indicated by relatively small changes of the external magnetic field required to elongate ferrogel up to the same length until the fifth day. The stability of the ferrogels can also be studied by observing the hysteresis loop of elongation as shown in **Figure 8**. Ferrogels fabricated by more than four cycles are generally unstable, implying that there is a reduction of water in the ferrogels during F-T cycle processes. This result is consistent with the previous papers [34, 49]. Through the F-T cycles, crystallites will be formed and act as the cross-linking points in the polymer matrix. The amount and size of these crystallites depend on the number of F-T cycles, as well as composition and concentration of the initial solution [15].

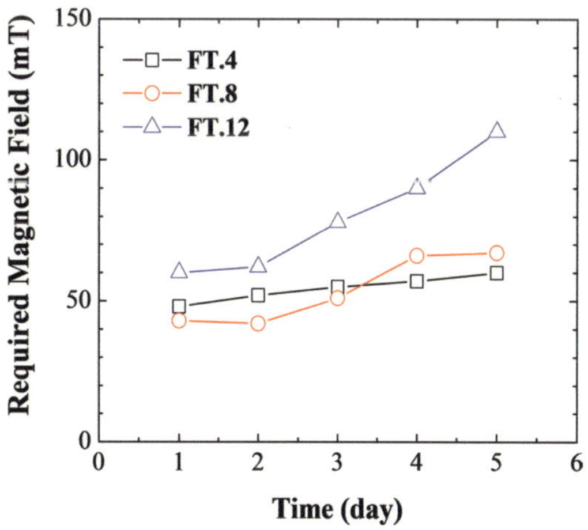

Figure 7. Required magnetic field to elongate ferrogel up to 1 mm as a function of time for ferrogels with a different number of F-T cycles.

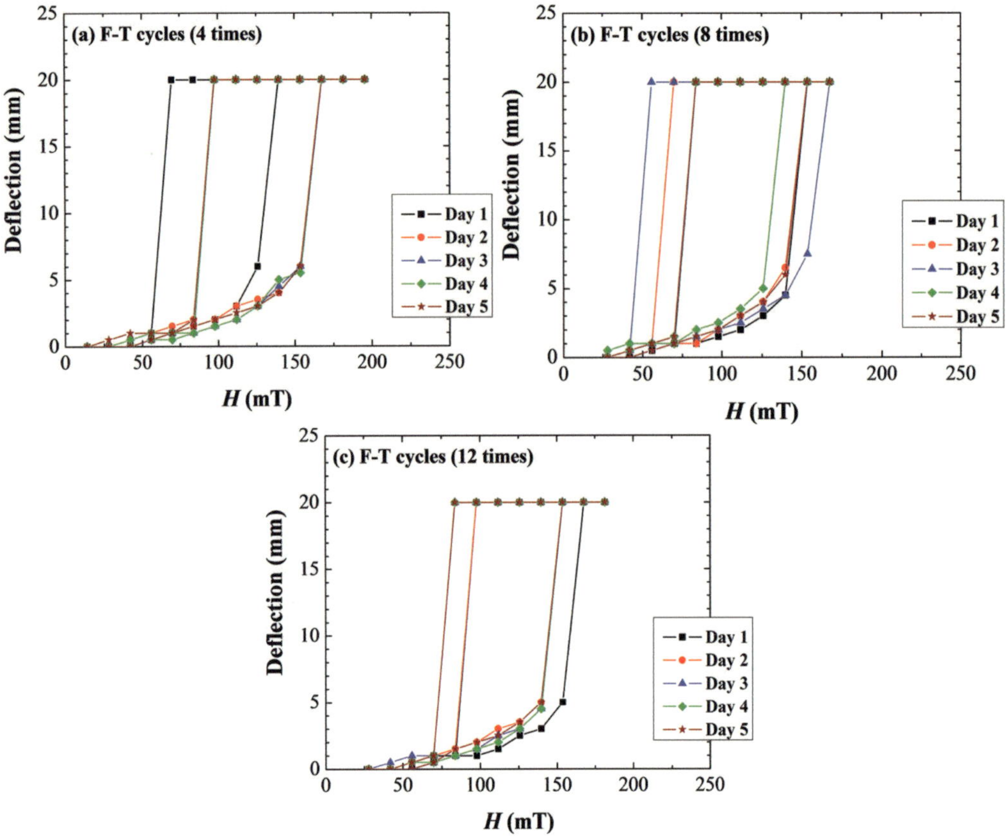

Figure 8. Hysteresis curves of the deflection behavior of ferrogels with the number of F-T cycles for: (a) 4, (b) 8, and (c) 12 times.

8. Structural and magnetic properties of Ferrogels

In addition to the substantial biomedical and biomechanical applications of PVA hydrogels and ferrogels, respectively, a basic study of structural and dynamical properties of ferrogel has to be investigated in detail. Structural studies using small-angle X-ray scattering (SAXS) measurement of PVA hydrogel and ferrogel have been reported by Puspitasari et al. [62] and Sunaryono et al. [48], respectively. Puspitasari et al. have confirmed that the crystallization of PVA hydrogel has a radius of approximately 2.9–3.3 nm and an average distance between polymer crystallites of 15–17.5 nm [62]. This result is consistent with the recent paper [48]. Moreover, Sunaryono et al. have illustrated the size distribution of Fe_3O_4 nanoparticles in the PVA hydrogel obtained by F-T cyclic process [48]. They have found that there are so-called primary particles (approximately 3 nm) and secondary particles as well as the clusters of magnetic nanoparticles in ferrogel observed by the synchrotron radiation (SAXS technique) with global fitting analysis data. The cluster size of the Fe_3O_4 in the ferrogel system was observed to be significantly reduced with decreasing concentration of the magnetic nanoparticles.

Ferrogel has potential application for an artificial muscle or a soft actuator due to the combined properties of good elasticity and flexibility from PVA hydrogel and specific magnetic behavior from the magnetic particles. Ramanujan et al. [40] have proposed two possible approaches of an artificial finger synthesized from PVA hydrogel and microsized iron oxide. First, they found that the deflection of ferrogel can be controlled by adjusting the concentration of magnetic particles. The second one is by coating manipulation of ferrogel. They demonstrated a finger-like motion based on instantaneous elongation and defection under external magnetic field.

As mentioned previously, in order to apply the ferrogel as an artificial tissue, one should understand the behavior of magnetoelastic properties. Based on numerous references [40, 48], the magnetoelasticity of ferrogel basically depends on the particle size and concentration of the magnetic particles in the polymeric matrix. Due to the particle size effect, the magnetization of microsized Fe_3O_4 particles in the ferrogel is generally higher than that of the nano-sized one. This may affect the threshold value of the magnetic field which is the minimum magnetic field required to start a large and instantaneous elongation or deflection of ferrogel. **Figure 9** shows the dependence of Fe_3O_4 concentration on the threshold magnetic field for both elongation and deflection of ferrogels with micro- and nano-sized Fe_3O_4 particles. It is found that the threshold magnetic field tends to decrease with increasing concentration of Fe_3O_4 particles. This result is consistent with the former report [40]. This result implies that the ferrogels are more sensitive to the external magnetic field with the increase of Fe_3O_4 concentration. **Figure 9** reflects the magnetic response for both variation of ferrogels, in which the ferrogel with microsized Fe_3O_4 particles has smaller threshold value than the ferrogel with nano-sized Fe_3O_4 particles as a consequence of the higher magnetization.

Figure 10 displays the Fe_3O_4 concentration dependence of elongation and deflection behaviors for both ferrogels with micro- and nano-sized Fe_3O_4 particles as fillers. It indicates that ferrogel with microsized filler is more sensitive to deform ferrogel under external magnetic field. This result can be explained by the lower threshold value of the magnetic field as illustrated in **Figure 9**. **Figure 11** shows the magnetoelastic hysteresis loops for ferrogels with different

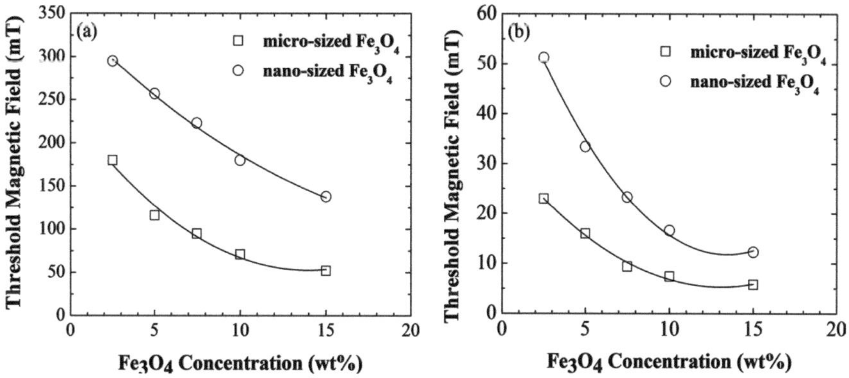

Figure 9. The threshold of magnetic field versus micro- and nano-sized Fe_3O_4 fillers for ferrogels during (a) elongation and (b) deflection. The solid lines are for the eye guidance.

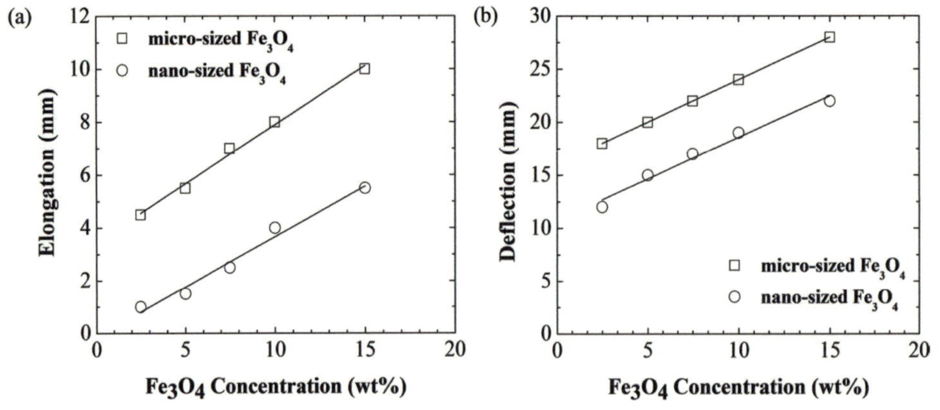

Figure 10. (a) Elongation and (b) deflection of ferrogels under 329 mT versus the concentration of micro- and nano-sized Fe_3O_4 particles as fillers. The solid lines are for the eye guidance.

concentration of microsized Fe_3O_4 particles from 2.5 to 15%. The hysteresis loops are found to be narrower and smaller with decreasing magnetic filler concentration. This is attributed to the different magnetic response of ferrogel to be deformed and returned to its original length and position. This behavior is also associated with a magnetic remnant of the ferrogels. A higher magnetic concentration leads to a higher ferrogel ability to deform even under a low external magnetic field. Moreover, a wider hysteresis loop for ferrogel with microsized filler was observed, indicating a stronger magnetic saturation. This is in a good agreement with the previous paper [27] that the ferrogel with large particle size has the best magnetosensitive effect, so it can be applied for drug release system.

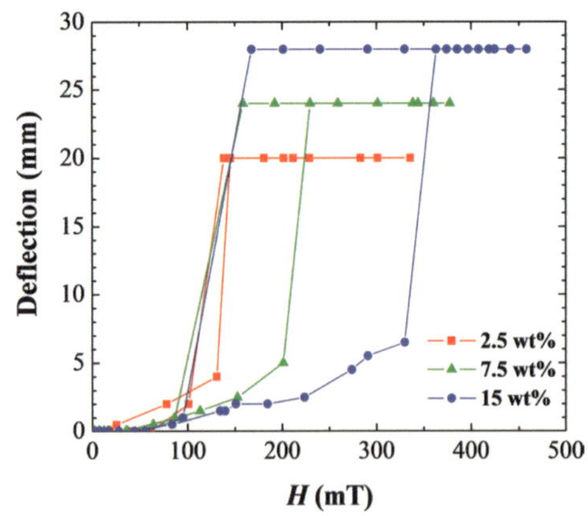

Figure 11. Hysteresis loops of the deflection versus electric current (proportional to the magnetic field) for ferrogels with different concentrations of Fe_3O_4 microparticles: 2.5 wt%, 7.5 wt%, and 15 wt%.

9. Conclusions

PVA hydrogel and ferrogels with Fe_3O_4 micro- and nano-sized particles as fillers have been successfully prepared by freezing-thawing (F-T) cyclic method. We can conclude the chapter as the following:

- In the biomedical application, a study of hydrogen (water molecules) diffusion behavior in the PVA hydrogel by analyzing the ADC value can be used as a parameter of brain tumor grading. The b-value of 1000 s/mm^2 and higher providing a better image quality and contrast is recommended for brain tumor grading.
- The time dependence of the elongation and deflection curves as a function of PVA concentration, particle concentration, and a number of F-T cycles can be used to determine the durability and performance of the ferrogel under certain external magnetic fields. It has been suggested that ferrogel with PVA and water ratio of 23:100 and four times F-T cycles, respectively, has the best elastic properties. Ferrogel fabricated by a F-T cyclic process has the best magnetoelastic response when it has a relatively large magnetic particle size as the filler with a concentration of 10–15 wt% in the PVA hydrogel.

Acknowledgements

This chapter is based on research funded by several schemes of research grants, provided by LPPM ITS, DP2M—Ministry of National Education, and DRPM—Ministry of Research, Technology and Higher Education, Republic of Indonesia, 2006–2016.

Author details

Malik Anjelh Baqiya[1], Ahmad Taufiq[2], Sunaryono[2], Munaji[3], Dita Puspita Sari[1], Yanurita Dwihapsari[1] and Darminto[1]*

*Address all correspondence to: darminto@physics.its.ac.id

1 Department of Physics, Faculty of Mathematics and Natural Sciences, Institut Teknologi Sepuluh Nopember (ITS), Kampus ITS Keputih Sukolilo, Surabaya, Indonesia

2 Department of Physics, Faculty of Mathematics and Natural Sciences, Universitas Negeri Malang, Malang, Indonesia

3 Faculty of Engineering, Universitas Muhammadiyah Ponorogo, Ponorogo, Indonesia

References

[1] Ullah F, Othman MBH, Javed F, Ahmad Z, Akil HM. Materials Science and Engineering: C. 2015;**57**:414

[2] Caló E, Khutoryanskiy VV. European Polymer Journal. 2015;**65**:252

[3] Gyles DA, Castro LD, Silva JOC, Ribeiro-Costa RM. European Polymer Journal. 2017; **88**:373

[4] Hamidi M, Azadi A, Rafiei P. Advanced Drug Delivery Reviews. 2008;**60**:1638

[5] Priya James H, John R, Alex A, Anoop KR. Acta Pharmaceutica Sinica B. 2014;**4**:120

[6] Bastiancich C, Danhier P, Préat V, Danhier F. Journal of Controlled Release. 2016;**243**:29

[7] Hennink WE, van Nostrum CF. Advanced Drug Delivery Reviews. 2012;**64**(Supplement, 223)

[8] Akhtar MF, Hanif M, Ranjha NM. Saudi Pharmaceutical Journal. 2016;**24**(5):554

[9] Ahmed EM. Journal of Advanced Research. 2015;**6**:105

[10] Qiu Y, Park K. Advanced Drug Delivery Reviews. 2012;**64**(Supplement, 49)

[11] Koetting MC, Peters JT, Steichen SD, Peppas NA. Materials Science and Engineering: R: Reports. 2015;**93**:1

[12] Schwall CT, Banerjee IA. Materials. 2009;**2**:577

[13] Hoare TR, Kohane DS. Polymer. 2008;**49**:1993

[14] Stauffer SR, Peppast NA. Polymer. 1992;**33**:3932

[15] Hassan CM, Peppas NA. Biopolymers PVA Hydrogels, Anionic Polymerisation Nanocomposites. Berlin Heidelberg, Berlin, Heidelberg: Springer; 2000. p. 37

[16] Ricciardi R, Auriemma F, De Rosa C, Lauprêtre F. Macromolecules. 2004;**37**:1921

[17] Ricciardi R, Auriemma F, Gaillet C, De Rosa C, Lauprêtre F. Macromolecules. 2004;**37**:9510

[18] Hernandez R, Lopez D, Mijangos C, Guenet J-M. Polymer. 2002;**43**:5661

[19] Hernández R, Sarafian A, López D, Mijangos C. Polymer. 2004;**45**:5543

[20] Park J-S, Park J-W, Ruckenstein E. Journal of Applied Polymer Science. 2001;**82**:1816

[21] Li Y, Hu Z, Chen Y. Journal of Applied Polymer Science. 1997;**63**:1173

[22] Gupta S, Webster TJ, Sinha A. Journal of Materials Science: Materials in Medicine. 2011; **22**:1763

[23] Siddhi G, Sudipta G, Arvind S. Biomedical Materials. 2012;**7**, 015006

[24] Fumio U, Hiroshi Y, Kumiko N, Sachihiko N, Kenji S, Yasunori M. International Journal of Pharmaceutics. 1990;**58**:135

[25] Jiang S, Liu S, Feng W. Journal of the Mechanical Behavior of Biomedical Materials. 2011;**4**:1228

[26] Armas AF, Gonzalez JS, Maiolo AS, Hoppe CE, Alvarez VA. Procedia Materials Science. 2012;**1**:483

[27] Liu T-Y, Hu S-H, Liu T-Y, Liu D-M, Chen S-Y. Langmuir. 2006;**22**:5974

[28] Liu T-Y, Hu S-H, Liu K-H, Liu D-M, Chen S-Y. Journal of Controlled Release. 2008;**126**:228

[29] Liu Y, Vrana NE, Cahill PA, McGuinness GB. Journal of Biomedical Materials Research Part B: Applied Biomaterials. 2009;**90B**:492

[30] Vrana NE, O'Grady A, Kay E, Cahill PA, McGuinness GB. Journal of Tissue Engineering and Regenerative Medicine. 2009;**3**:567

[31] Jiang S, Su Z, Wang X, Liu S, Yu Y. Materials Science and Engineering: C. 2013;**33**:3768

[32] Yanagawa F, Sugiura S, Kanamori T. Regenerative Therapy. 2016;**3**:45

[33] Forte AE, Galvan S, Manieri F, Rodriguez F, Baena Y, Dini D. Materials & Design. 2016;**112**:227

[34] Gonzalez JS, Hoppe CE, Muraca D, Sánchez FH, Alvarez VA. Colloid and Polymer Science. 2011;**289**:1839

[35] Sivudu KS, Rhee KY. Colloids and Surfaces A: Physicochemical and Engineering Aspects. 2009;**349**:29

[36] Martínez H, D'Onofrio L, González G, in LACAME 2012: Proceedings of the 13th Latin American Conference on the Applications of the Mössbauer Effect, (LACAME 2012) held in Medellín, Colombia, November 11-16, 2012. edited by C. A. B. Meneses et al. Netherlands, Dordrecht: Springer; 2014. p. 93

[37] Torre B, Bertoni G, Fragouli D, Falqui A, Salerno M, Diaspro A, Cingolani R, Athanassiou A. Scientific Reports. 2011;**1**:202

[38] Weeber R, Kantorovich S, Holm C. The Journal of Chemical Physics. 2015;**143**, 154901

[39] Zrínyi M, Barsi L, Büki A. Polymer Gels and Networks. 1997;**5**:415

[40] Ramanujan RV, Lao LL. Smart Materials and Structures. 2006;**15**:952

[41] Reséndiz-Hernández PJ, Rodríguez-Fernández OS, García-Cerda LA. Journal of Magnetism and Magnetic Materials. 2008;**320**, e373

[42] Hernandez R, Sacristan J, Nogales A, Fernandez M, Ezquerra TA, Mijangos C. Soft Matter. 2010;**6**:3910

[43] Goiti E, Salinas MM, Arias G, Puglia D, Kenny JM, Mijangos C. Polymer Degradation and Stability. 2007;**92**:2198

[44] Zubarev A. Physica A: Statistical Mechanics and its Applications. 2013;**392**:4824

[45] Zubarev AY, Elkady AS. Physica A: Statistical Mechanics and its Applications. 2014;**413**:400

[46] Zubarev AY, Borin DY. Journal of Magnetism and Magnetic Materials. 2015;**377**:373

[47] Moscoso-Londoño O, Gonzalez JS, Muraca D, Hoppe CE, Alvarez VA, López-Quintela A, Socolovsky LM, Pirota KR. European Polymer Journal. 2013;**49**:279

[48] Sunaryono A, Taufiq EGR, Putra A, Okazawa I, Watanabe N, Kojima S, Rugmai S, Soontaranon M, Zainuri, Triwikantoro S, Pratapa, Darminto. Nano. 2016;**11**, 1650027

[49] Sunaryono A, Taufiq, Munaji B, Indarto, Triwikantoro M, Zainuri, Darminto. AIP Conference Proceedings. 2013;**1555**:53

[50] Dwihapsari Y, Sari DP, Darminto. AIP Conference Proceedings. 2012;**1454**:53

[51] Sari DP, Kristanto SA, Wahyudi RE, Dwihapsari Y, Darminto: In Instrumentation, Communications, Information Technology, and Biomedical Engineering (ICICI-BME), 2013 3rd International Conference on 2013. p. 302

[52] Triwikantoro MA, Baqiya T, Heriyanto, Mashuri, Darminto. Journal of Superconductivity and Novel Magnetism. 2017;**30**:555

[53] Baqiya MA, Taufiq A, Sunaryono K, Ayun M, Zainuri S, Pratapa, Triwikantoro, Darminto: In: Magnetic Spinels- Synthesis, Properties and Applications, edited by Seehra MS. Rijeka: InTech; 2017. p. Ch. 11

[54] Moritani T, Ekholm S, Westesson P-L. Diffusion-Weighted MR Imaging of the Brain. 2nd ed. New York: Springer; 2009

[55] Seo HS, Chang K-H, Na DG, Kwon BJ, Lee DH. American Journal of Neuroradiology. 2008;**29**:458

[56] Hassan CM, Ward JH, Peppas NA. Polymer. 2000;**41**:6729

[57] Amsden B. Macromolecules. 1998;**31**:8382

[58] Fromageau J, Gennisson J, Schmitt C, Maurice R, Mongrain R, Cloutier G. IEEE Transactions on Ultrasonics, Ferroelectrics, and Frequency Control. 2007;**54**:498

[59] Zrínyi M, Barsi L, Szabó D, Kilian H-G. The Journal of Chemical Physics. 1997;**106**:5685

[60] Snyder RL, Nguyen VQ, Ramanujan RV. Acta Materialia. 2010;**58**:5620

[61] Morozov K, Shliomis M, Yamaguchi H. Physical Review E. 2009;**79**:040801

[62] Puspitasari T, Raja KML, Pangerteni DS, Patriati A, Putra EGR. Procedia Chemistry. 2012;**4**:186

Chapter 9

Hyaluronic-Based Antibacterial Hydrogel Coating for Implantable Biomaterials in Orthopedics and Trauma: From Basic Research to Clinical Applications

Giammona Gaetano, Pitarresi Giuseppe,
Palumbo Fabio Salvatore, Maraldi Susanna,
Scarponi Sara and Romanò Carlo Luca

Additional information is available at the end of the chapter

http://dx.doi.org/10.5772/intechopen.73203

Abstract

Bacterial colonization of implanted biomaterials remains one of the most challenging complications in orthopedics and trauma surgery, with extremely high social and economic costs. Antibacterial coating of implants has been advocated by many experts as a possible solution to reduce the burden of implant-related infection and several different solutions have been proposed in the last decades. However, while most of the investigated technologies have shown their efficacy *in vitro* and/or *in vivo*, only few were able to reach the market, due to clinical, industrial, economic and regulatory issues. Hyaluronic acid composites have been previously shown to possess antifouling capabilities and have been used in various clinical settings to reduce bacterial adhesion and mitigate biofilm-related infections. Recently, a fast-resorbable, hyaluronic-based hydrogel coating was developed to protect implanted biomaterials in orthopedics, trauma and maxillofacial surgery. Preclinical and clinical testing did show the safety and efficacy of the device that can be intraoperatively loaded with one or more antibiotics and directly applied by the surgeon to the implant surface, at the time of surgery. Here, we review the current evidence concerning this very first antibacterial coating of implants and outline the economic impact of the possible large-scale application of this technology.

Keywords: coating, hydrogel, hyaluronic acid, DAC, infection, implant, orthopedic, trauma, prosthesis, prevention

1. Introduction

Up to 80% of human bacterial infections are biofilm-related, according to the U.S. National Institutes of Health [1]. Among these, implant-related infections in orthopedics and trauma still have a tremendous impact [2]. In fact, periprosthetic joint infection (PJI) (**Figure 1**) is among the first reasons for implant failure [3], posing challenging diagnostic and therapeutic dilemmas [4] and with high economic and social costs [5–7].

Similarly, surgical site infections after osteosynthesis, with a reported incidence ranging from 3.9 to 10% for closed fractures [8–11] and even more after open fractures [12], are associated with high morbidity and possible mortality raise [9] and elevated costs [13].

Whenever a biomaterial is implanted, a competition between host and bacterial cells occurs for surface colonization. In the event of bacterial adhesion to an implant, immediate biofilm formation starts, making the bacteria extremely resistant to host's defense mechanisms and to antimicrobials [14–16]. According to recent evidence, fully formed biofilm can be found few hours after the first bacterial adhesion [17]; thus, the destiny of an implant is decided at the very time of surgery.

To reduce or prevent bacterial adhesion and biofilm formation, a number of different antimicrobial finishing or coatings of implants are under study [18]. However, their clinical application appears particularly challenging, due to the many requirements they need to fulfill [19].

Hyaluronic acid (HA) is mucopolysaccharide, occurring naturally in mammals. It is abundant in skin and in connective tissues, being one of the main components of extracellular matrices. HA has several clinical applications in dermatology, esthetic surgery, dentistry, urology, orthopedics and ophthalmology [20]. In fact, due to its high biocompatibility, and nonimmunogenicity, hyaluronic acid is considered as an ideal biomaterial for medical and pharmaceutical applications [21, 22].

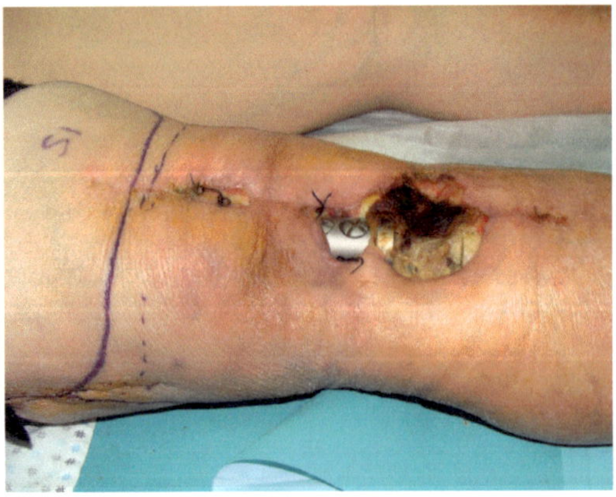

Figure 1. Infected, exposed, knee prosthesis in a 60-year-old woman. Approximately one million joint replacements are performed annually in Europe, and infection is currently among the first three most common reasons for failure of implants. Septic complications are associated with prolonged and complex medical and surgical treatments, often leading to implant removal. Poor functional results, possible infection recurrence, risk of amputation and increased mortality rate are all well known and feared consequences of periprosthetic and implant-related infections. Direct costs of treatment of periprosthetic infection exceeds 100,000 euros, per case, according to a recent analysis [7].

Local application of hyaluronic-based compounds has been demonstrated to be protective against various infectious agents, depending on HA concentration and molecular weight; furthermore, HA's ability to reduce bacterial adhesion and biofilm formation has been recently reported [23].

High biocompatibility, safety profile and antiadhesive properties make HA and its composites a possible non-antibiotic option to reduce the impact of biofilm-related infections in various clinical settings. However, the use of HA in its pure form as an antibacterial coating does not appear suitable, due to its rapid degradation by hyaluronidases, enzymes naturally occurring in the human and animal body. Furthermore, due to its high hydrophilicity, a coating produced with a hydrogel of HA alone would not have sufficient mechanical stability when a prosthesis is implanted in the body, which is an essentially water-based environment.

To overcome these limits, a combination of HA with another biocompatible and biodegradable polymer, polylactic acid (PLA), was investigated [24]. In fact, PLA is a synthetic polyester, approved in the U.S.A. by the Food and Drug Administration (FDA) and widely used for orthopedic implants [25]. PLA unlike HA, shows a hydrophobic character; therefore, its presence could be exploited to control in appropriate way the hydrophilic and mechanical properties of a hydrogel based on HA, thus slowing down the susceptibility to hydrolysis.

Here, after an overview of the antiadhesive and antibiofilm properties of HA, we summarize the development of a CE-marked, patented hydrogel coating, based on HA grafted to PLA (DAC®, "Defensive Antibacterial Coating," Novagenit Srl, Mezzolombardo, Italy). Some of the most relevant preclinical and clinical results that made this device the very first resorbable antibacterial coating for large scale clinical applications in orthopedic, trauma, dentistry and maxillofacial implants are also briefly reported.

2. Antiadhesive and antibiofilm properties of HA

Pavesio et al. [26] were probably the first to describe HA nonfouling properties and its ability to resist bacterial adhesion, with particular reference to *Staphylococcus epidermidis* [27], proposing coated polymeric medical devices to reduce implant-related infections. In particular, a hydrophilic HA overlayer, linked to the surface of polymethylmethacrylate intraocular lenses (IOLs), was shown to be able to significantly reduce the adhesion of *Staphylococcus epidermidis* to the implant surface [28].

In line with this observation, Kadry and coworkers, reported the ability of hyaluronan to reduce bacterial adhesion to IOLs of a *S. epidermidis* wild strain [29]; based on these findings, the authors proposed the use of HA as an antiadhesive, adjuvant therapy, in combination with antibiotics in irrigating solutions for bacterial ocular infections.

More recently, Drago et al. reported on the *in vitro* antiadhesive and antibiofilm activity of HA toward bacterial species commonly isolated from respiratory infections [30]. In this experimental study, HA was shown to be able to reduce bacterial adhesion to a cellular substrate in a concentration-dependent manner. The antibiofilm action, exerted by HA in ear, nose and throat districts, has been recently reviewed [31]. The authors conclude that "its efficacy in treating rhinosinusitis, whether or not associated with polyposis, is well documented, as well as results from its effects on mucociliary clearance, free radical production and mucosal repair."

HA has also been reported to exert bacteriostatic, dose-dependent effect on different planktonic microorganisms [32, 33]. Radaeva et al. showed the inhibiting activity of HA with respect to some *Pseudomonas* species [34], while Ardizzoni and coworkers [23] investigated the effects of HA on 15 ATCC bacterial strains, representative of clinically relevant bacterial and fungal species. According to their results, different microbial species and strains are differently affected by HA. In particular, staphylococci, enterococci, *Streptococcus mutans*, two *Escherichia coli* strains, *Pseudomonas aeruginosa, Candida glabrata* and *C. parapsilosis* showed a dose-dependent growth inhibition, while no HA effects were detected in *E. coli* ATCC 13768 and *C. albicans*, and *S. sanguinis* was favored by the highest HA dose.

Carlson and coworkers [33] compared the potential bacteriostatic effect of collagen type I, hyaluronic acid, hydroxyapatite, polylactic acid and polyglycolic acid on some of the most common orthopedic bacterial pathogens (*S. aureus, S. epidermidis, β-hemolytic Streptococcus* and *Pseudomonas aeruginosa*): HA had the most significant bacteriostatic properties on the studied organisms. Similarly, Pirnazar et al. [32] did demonstrate the bacteriostatic effect of HA in different concentrations and molecular weight on oral and nonoral microorganisms (*Staphylococcus aureus, Propionibacterium acnes, Actinobacillus actinomycetemcomitans, Prevotella oris* and *Porphyromonas gingivalis*). The authors concluded that the clinical application of hyaluronan in the form of membranes, gels or sponges may reduce bacterial contamination of the surgical wound, thereby lessening the risk of postsurgical infection and promoting more predictable regeneration.

Concerning orthopedic applications, Harris and Richards [35] showed how coating titanium with sodium hyaluronate significantly decreased the density of *S. aureus* adhering to the surfaces and proposed its potential use to protect osteosynthesis, orthopedic or dental implants.

In a recent review, focused on the use of polysaccharide-based coatings to prevent biofilm formation, hyaluronic acid was discussed as one of the most promising [36]; displaying hydrophilic characteristics, this coating was in fact reported to reduce adhesion of *S. aureus, S. epidermidis* and *E. coli* by several orders of magnitude compared to unmodified surfaces.

3. Clinical applications of HA to prevent bacterial adhesion

Several clinical local applications of HA to reduce the impact of biofilm-related infections have been reported with favorable results and no adverse events [37].

Torretta et al. [38] recently described topical administration of hyaluronic acid in children with recurrent or chronic middle ear inflammations and chronic adenoiditis.

Other studies have documented the positive effect of topical HA in chronic urinary tract infections (UTI). At variance with current antibiotic treatments, aimed at eradicating pathogens, HA local administration targets bacterial adherence to the bladder mucosa [39–42]. Damiano et al., in a prospective, randomized, double-blind, placebo-controlled study, showed a significant reduction of 77% ($P < 0.0002$) in the UTI rate per year in HA-treated patients, compared to controls. Moreover, mean time to UTI recurrence was significantly prolonged (185.2 ± 78.7

http://dx.doi.org/10.5772/intechopen.73203

vs. 52.7 ± 33.4 days, $P < 0.001$) after HA treatment, compared with placebo [43]. No adverse events were reported. A recent multicenter European study confirmed the efficacy of intravescical administration of combined HA and chondroitin sulfate (CS) for the treatment of female recurrent urinary tract infections [44].

In dentistry, the effect of the application of HA-containing gels in early wound healing after scaling and root planing (SRP) on clinical variables, subgingival bacteria and local immune response was investigated [45, 46]. Eick et al. [47] reported on 34 individuals affected by chronic periodontitis and treated with full-mouth SRP; in the test group (n = 17), a 0.8% hyaluronan-containing gel was introduced into all periodontal pockets during SRP and a 0.2% HA gel was applied by the patients onto the gingival margin twice daily during the following 2 weeks, while the control group (n = 17) was treated with SRP only; no placebo was used. Probing depth (PD) and clinical attachment level (CAL) were recorded at baseline and after 3 and 6 months, and subgingival plaque and sulcus fluid samples were taken for microbiologic and biochemical analysis. The changes in PD and the reduction of the number of pockets with PD ≥ 5 mm were significantly higher in the test group after 3 (P = 0.014 and 0.021) and 6 (P = 0.046 and 0.045) months. Six months after SRP, the counts of *Treponema denticola* were significantly reduced in both groups (both P = 0.043), as were those of Campylobacter rectus in the test group only (P = 0.028). *Prevotella intermedia* and *Porphyromonas gingivalis* increased in the control group. No adverse effects of HA were observed during the study.

4. Synthesis of DAC® HA-g-PLA hydrogel coating

Composed of covalently linked hyaluronan and poly-D,L-lactide, the "Defensive Antibacterial Coating" (DAC®, Novagenit Srl, Mezzolombardo, Italy) was specifically developed in order to protect implanted biomaterials used in orthopedics, traumatology, dentistry and maxillofacial surgery from bacterial colonization [24, 48] (**Figure 2**).

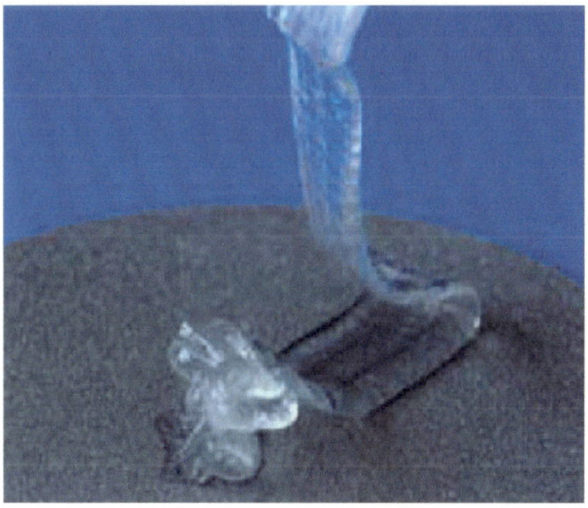

Figure 2. DAC® HA-g-PLA, fast-resorbable, hydrogel coating. Composed of covalently linked hyaluronan and poly-D, L-lactide, the "Defensive Antibacterial Coating" (DAC®, Novagenit Srl, Mezzolombardo, Italy) is the first antibacterial-coating cleared for clinical use in orthopedics, trauma, dentistry and maxillofacial surgery in Europe.

Preparation of the hydrogel was performed according to a patented procedure [49]. In particular, HA-g-PLA copolymer was dispersed in an appropriate volume of twice distilled water, and the mixture was stirred vigorously at the vortex to obtain a gelatinous and transparent hydrogel with a polymer concentration between 3% (w/v) and 10% (w/v).

The synthesis of HA-g-PLA copolymer was performed as previously reported [50, 51] (**Figure 3**). Briefly, a low weight–average molecular weight HA (HALMW) was made soluble in organic solvents by transformation to its tetrabutylammonium (TBA) salt. The synthesis of the N-hydroxysuccinimide (NHS) derivative of PLA (i.e., PLA-NHS) was performed as reported elsewhere [52]. In particular, 2.4 g of PLA was dissolved in 30 ml of anhydrous dichloromethane with an excess of DCC and NHS for 24 h at room temperature, and then the solution was precipitated in ethanol and the recovered solid was dried under vacuum. 1H NMR of PLA-NHS ($CDCl_3$) showed: δ 1.5 and δ 1.6 (d, 3H, —O—CO—CH(CH_3)—OH; d, 3H, —O—CO—CH(CH_3)—O—),

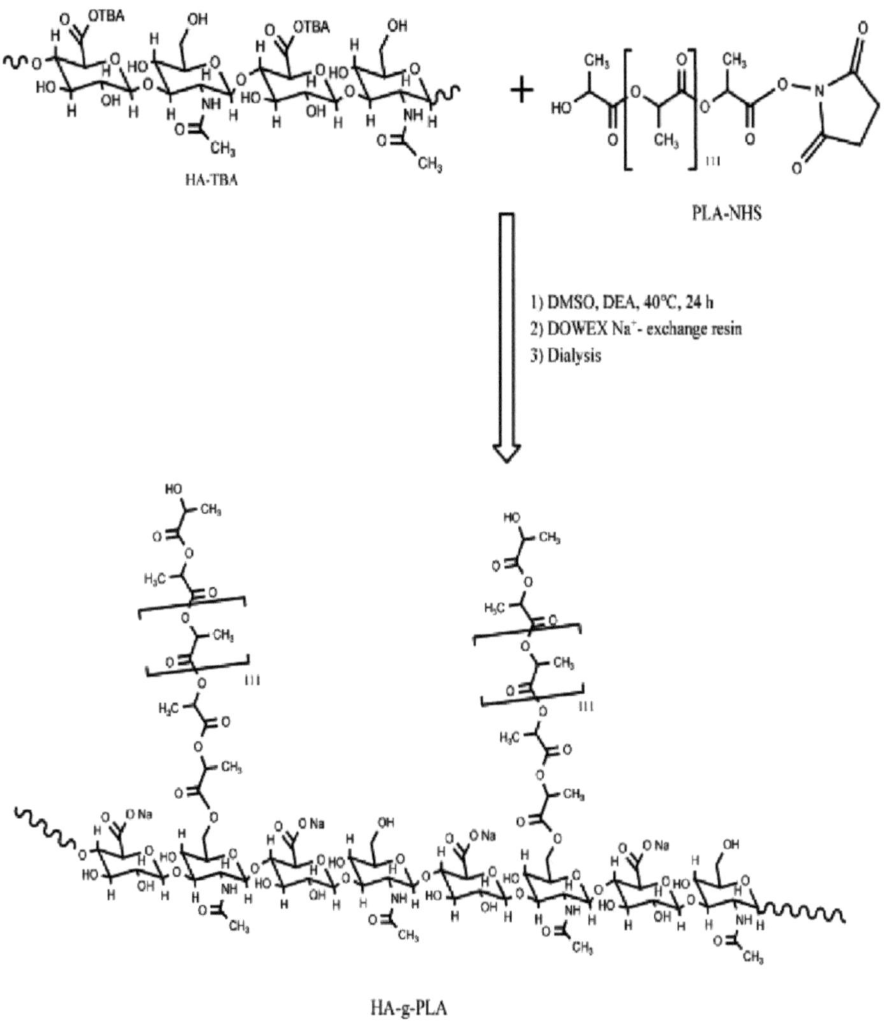

Figure 3. Principal steps in the synthesis of HA-g-PLA copolymer.

http://dx.doi.org/10.5772/intechopen.73203

δ 2.80 (m, 4H, —OC—CH_2—CH_2—CO—); δ 4.3 and δ 5.2 (m, 1H, —O—CO—**CH**(CH_3)—OH; m, 1H, —O—CO—**CH**(CH_3)—O). The synthesis of HA-g-PLA copolymer was carried out as follows: 600 mg of HA-TBA was dissolved in 48 ml of anhydrous dimethyl sulfoxide (DMSO) and then 576 μl of DEA, as a catalyst, was added. A suitable amount of PLA-NHS (dissolved in 6 ml of anhydrous DMSO) was added according to $X = 1$, X being equal to moles of PLA-NHS/moles of HA repeating units. The PLA-NHS solution was added drop by drop to the HA-TBA solution in about 1 h. The reaction was carried out under argon at 40°C for 24 h. After this time, the TBA was exchanged with Na^+ using a Dowex 50 W × 8-200 resin, and then the eluate was dialysed against distilled water, by using spectra/por tubing with a cutoff of 14,000 Da and then freeze-dried. The sample has been characterized by FT-IR and 1H NMR analyses. FT-IR spectrum (KBr) of HA-g-PLA showed a broad band centered at 3450 cm^{-1} (ν as OH + ν as NH of HA), bands at 1757 (ν as COO of PLA), 1623 (amide I of HA), 1456 (δ as CH_3 of PLA), 1382 (δ as CH_3 of PLA), 1189 (ν as C—O—C ester group of PLA), 1089, 1048 (ν C—O alcoholic and ether of HA) cm^{-1}. 1H NMR of HA-g-PLA (DMSO-d_6/D_2O 90:10) spectrum showed: δ 1.25 and δ 1.45 (2d, —O—CO—CH(CH_3)—O— of PLA); δ 1.85 (s, 3H, —NH—CO—CH_3 of HA) δ 5.1 ppm (m, —O—CO—CH(CH_3)— of PLA). The % degree of grafting (DG) has been calculated as: %DG = (moles PLA chains/moles of HA repeating units) × 100. The degree of grafting was determined by comparing the integral of the peaks at δ 1.25–1.45 attributed to protons of methyl groups of PLA with the integral of peaks at δ 1.85 attributed to protons of NHCO$\mathbf{CH_3}$ belonging to N-acetylglucosamine residue of HA and resulted to be 7 ± 1 mol%.

5. DAC® hydrogel *in vitro* activity

5.1. Cell compatibility assay

In vitro cell compatibility of DAC® HA-g-PLA hydrogel (polymer concentration 6%, w/v) was evaluated using human dermal fibroblasts. The viability of cells cultured in direct or indirect contact with HA-g-PLA hydrogel was comparable with that of the control well, showing that the hydrogel does not release in the culture medium substances that interfere with cell viability and they do not cause a decrease in the cell viability after direct contact with them [24]. Further *in vitro* and *in vivo* biocompatibility studies were performed on the DAC® hydrogel and on the DAC® kit, in accordance to ISO standards, all showing no cytotoxicity, genotoxicity, sensitization, irritation or intracutaneous reactivity, systemic toxicity (acute), subchronic toxicity or interference with bone or periimplant tissues.

Furthermore, as degradation of DAC® HA-g-PLA hydrogel occurs via deesterification of hyaluronic acid and polylactic acid, it gives raise exclusively to the starting macromolecules, whose degradation pathways in the human body are widely known and whose use as implantable class III medical devices is largely accepted and tested safe.

5.2. Antiadhesive and antibiofilm activity

Both the ability of the DAC® HA-g-PLA hydrogel to reduce bacterial adhesion and biofilm formation were extensively studied *in vitro*.

Reductions of adhered bacteria on sterile titanium discs, coated with DAC® hydrogel, equal to 86.8, 80.4, 74.6 and 66.7% vs. untreated discs were observed after 15, 30, 60 and 120 min of incubation, respectively [37]. In another experiment, the ability to dislodge previously adhered bacteria was investigated. Once again, the results showed that DAC® hydrogel treatment of discs reduced the amount of adhered bacteria in respect to control discs after 15, 30, 60 and 120 min by 84.0, 72.8, 72.3 and 64.3%, respectively [37].

Concerning more specifically the antibiofilm activity, DAC® hydrogel showed similar or superior *in vitro* activity, compared to various antibacterials and a synergistic activity when used in combination [48]. In one experimental setting, *S. epidermidis* and *S. aureus* were grown on chrome-cobalt devices in 6-wells polystyrene plates containing TSB for 24 h at 37°C. The plates were incubated at 37°C in ambient air, until a visible biofilm was obtained. Gentamycin and vancomycin were tested at a final concentration of 20 mg/mL. Similarly, when mixed with the hydrogel, 60 mg of gel powder was reconstituted with 1 mL of water for injections containing gentamicin or vancomycin at 20 mg/mL concentration. The amount of biofilm at each time was determined before hydrogel and antibiotic agents' addition and after 0.5, 1, 2, 4, 6, 24 and 48 h of incubation by a spectrophotometric assay. At each time point, both gentamicin and vancomycin showed only a partial inhibition of biofilm formation (ca. 30–40% for gentamicin; ca. 40–50% for vancomycin), with minor difference between the two studied microorganisms. On the other side, the hydrogel alone resulted in a significant reduction of biofilm of ca. 50%, in comparison to the untreated controls, while a combination of the hydrogel with either antibacterial coating resulted in a larger reduction of biofilm formation (approximately 75–80% in comparison with untreated controls).

Both these experimental studies show the ability of the DAC® hydrogel to significantly reduce bacterial adhesion and biofilm formation of common bacterial pathogens, thus potentially providing an effective protection of the implant; however, these data also point out how, in the clinical setting, in the absence of an adequate immune response from the host and/or of sufficient local levels of antibiotics, a passive antiadhesive coating [18] like HA can be overcome by the remaining bacteria in a time-dependent manner. For this reason, any passive antiadhesive coating of implants [53] should probably better be seen as a tool to reduce and delay bacterial adhesion and biofilm formation to a variable degree, also depending on the local environment, the contaminating bacterial species and initial bacterial load; this activity of the coating may represent a key additional advantage to the host's cells to win the competition with the microorganisms that may eventually be present. However, the known intrinsic limits of all passive coatings ground the idea of adding antibacterial agents to the protective hydrogel, in order to minimize the possibility for planktonic bacteria, which may eventually remain in the local environment, to colonize the implant at a second stage, when the coating has been hydrolyzed or covered by host's proteins.

5.3. Antibiotic release studies

Although designed as a "stand alone" product, the DAC® hydrogel was also tested concerning its ability to entrap and eventually release locally various antibacterial agents. As outlined above, the rationale for this combination lies in the pathogenesis of implant-related infections and on the specificities of passive protective coatings. In fact, biofilm formation is a multistep process that schematically includes the bacterial adhesion to a substrate, the subsequent immediate release of signals from adherent bacteria that triggers biofilm production and, finally, the biofilm construction and progressive consolidation. Acting mainly as a physical and antiadhesive barrier, DAC® hydrogel may reduce or prevent the first phase of the process, provided that the number of living bacteria is not too high and that they are not able to overcome or hydrolyze the hydrogel [54]; moreover, for an effective prevention of bacterial colonization of the coated implant, it is necessary that, while bacteria are in the more vulnerable planktonic state, they are completely removed or killed by the host's immune system and/or by the local chemicophysical environment. This is why, even in the presence of DAC® coating protection, systemic antibiotic prophylaxis is still to be considered necessary. In fact, if not eliminated, the remaining floating microorganisms may successfully colonize the implant once the protective coating has been hydrolyzed (a phenomenon that is expected to happen normally within 3 days from application for the DAC hydrogel) or after the implanted biomaterial has been covered by host's proteins (fibrin, fibronectin, etc.), which may also work as suitable for bacterial adhesion. In this scenario, the possibility to add also an antibacterial drug to the coating may further contribute to reduce the planktonic microorganisms, enhancing the overall protection offered by the DAC hydrogel (**Figure 4**).

To test the hydrogel ability to entrap antibacterial drugs, vancomycin and tobramycin had been originally chosen as examples of antibiotic molecules [24]. Both these antibiotics have been added to the hydrogel just before its use, a solution that offers several advantages. First of all, this allows to add the antibiotic when it is needed, thus avoiding the problems of shelf-life and any long-term compatibility with the hydrogel; secondly, in this way, it is possible to choose the optimal antibiotic in the specific case, taking into account the patient's specificity (e.g., known intolerances to specific antibacterial agents) or of the specific intervention; finally, the dosing of the antibacterial agent on a case-by-case basis can be decided by the clinician. The results of the *in vitro* study clearly showed how the investigated antibacterial hydrogel coating, applied on a titanium disc, at a concentration in polymer in the range 2–8% (w/v) and a concentration in drug equal to 1 or 2% (w/v) is able to release vancomycin or tobramycin, or of their association, for up to 72 h, with an amount of drug released that is hundred or thousand times higher than the minimum inhibitory concentration (MIC), in a time- and dose-dependent manner.

Similar results were founded by testing several other antibacterial compounds or their combinations, including vancomycin, teicoplanin, rifampicin, daptomycin, tigecycline, cefazolin, gentamicin, tobramycin, amikacin, meropenem, levofloxacin, etc. (cf. **Figure 5**), at concentrations ranging from 2 to 10% [48].

In summary, all examples reported above show that DAC® hydrogel is potentially able to entrap and release suitable quantities of antibacterial agents just after the implant of the coated prosthesis. The high initial burst effect of the released drug may ensure the most efficient action at the time that is critical for the destiny of the biomaterial. Moreover, during the entire drug release period, the antibiotic concentration released by the hydrogel remains greater than MIC, thus further ensuring effectiveness of the drug released in proximity to the prosthesis.

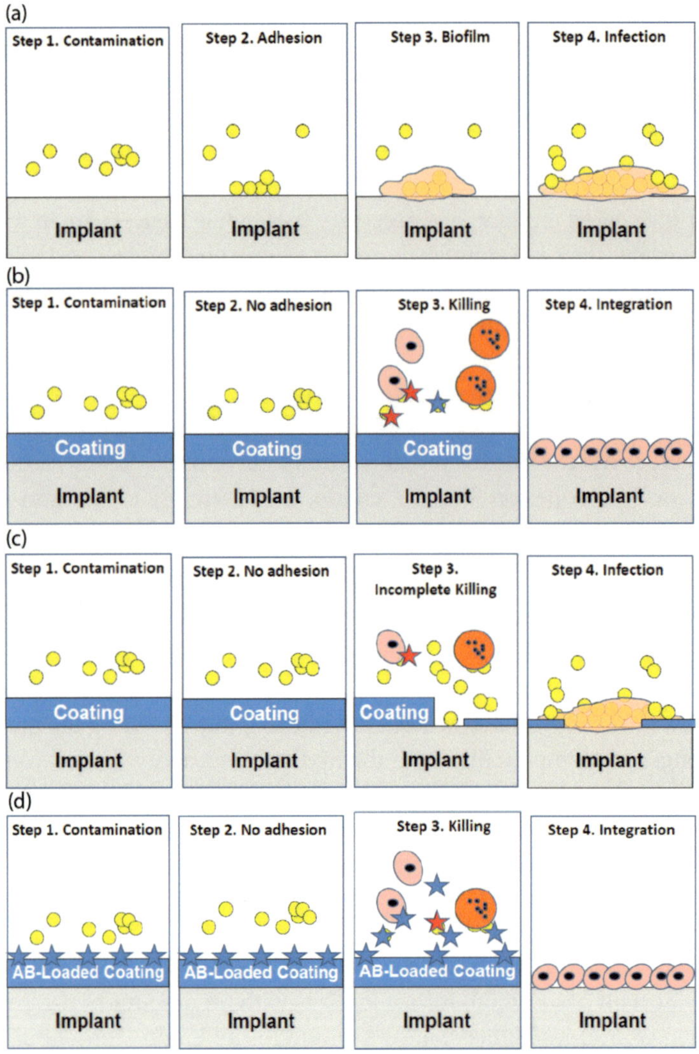

Figure 4. Rationale for intra-operative mixing of DAC® hydrogel coating with antibacterial agents. Schematic representation of different scenarios. (a) Noncoated implants may get colonized by biofilm-forming bacteria (yellow circles) and infection will develop. (b) Antiadhesive coating may reduce/prevent bacterial adhesion, while the immune system (orange circles and red stars) and the systemically administered antibiotics (blue star) kill planktonic microorganisms. (c) However, if bacterial load is large enough, or if immune response and local antibiotic levels are inadequate, surviving bacteria may eventually colonize the implant, once the coating has been hydrolyzed or covered by host's proteins. (d) To prevent this, the antibacterial hydrogel may be loaded, at the time of surgery, with antibiotic agents (blue stars) that may be locally released, contributing to eliminate all remaining planktonic bacteria.

5.4. DAC® hydrogel coating ability

For any device candidate to act as a coating of orthopedic and trauma biomaterials, mechanical adherence to the implant surface plays a key role. In particular, DAC® hydrogel has been designed to be spread manually at the time of surgery and to not interfere with the usual surgical techniques of press-fit insertion of an implant. The ability of DAC® hydrogel to completely cover even sand-blasted titanium surface and resist scraping has been confirmed by

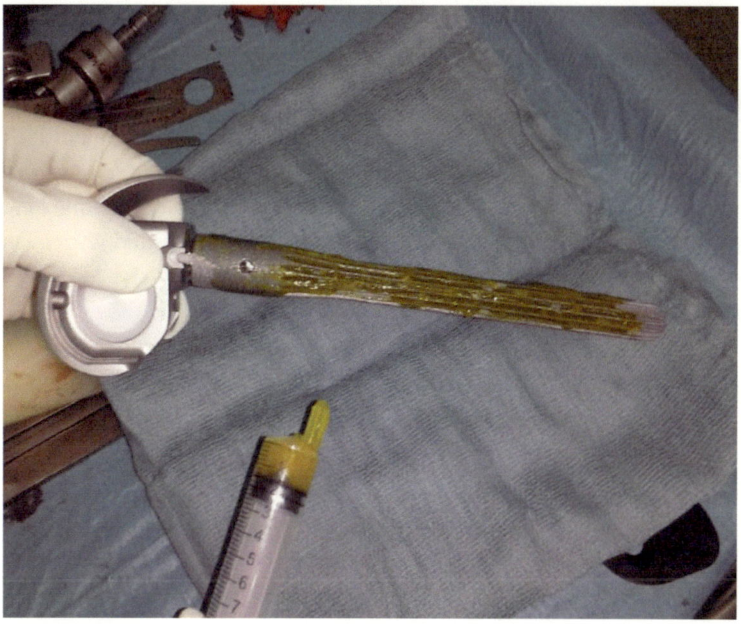

Figure 5. Tigecycline-loaded DAC® hydrogel coating, applied at surgery on a knee revision prosthesis. The hydrogel, which comes in a powder form, in a prefilled syringe, is designed to be reconstituted at the time of surgery with water for injection. The surgeon may decide to add a single or a combination of antibiotics to the water for injection, to further enhance implant protection.

scanning electron microscopy (SEM) analysis [37]. This is an important requirement in order to reduce the exposed surface of a biomaterial, thus creating a uniform coating of the surface and leaving no pores or cracks that could eventually be colonized by planktonic bacteria.

The resistance to scraping and declothing has also been tested in the animal model and in human femurs, simulating a press-fit insertion of a cementless implant [24, 48]. Both studies demonstrated the ability of the hydrogel coating to resist insertion, with 60% to more than 80% of the hydrogel remaining adherent to the entire implant surface, while the remainder being retrieved along the inner surface of the medullary canal.

6. DAC® hydrogel *in vivo* activity

Safety and efficacy of the DAC® hydrogel have been investigated in several animal studies.

Concerning efficacy, in an acute model of highly contaminated implant-related infection in the rabbit, Giavaresi and coworkers [55] found that a vancomycin-loaded DAC® coating was associated with local bacterial load reduction ranging from 72 to 99%, compared to uncoated controls.

In another large investigation in the rabbit model, Boot et al. [56] showed, at longer follow-up and without systemic antibiotic prophylaxis, the ability of vancomycin-loaded DAC®-coated implants to significantly resist infection, compared to uncoated controls. Both studies did also reveal the absence of local or systemic side effects. In line with this observation, a more recent paper, focused on the impact on bone healing and implant osteointegration, reported no detrimental effects of vancomycin-loaded DAC® or of DAC® alone [57].

Another study on a rat model of acutely infected osteosynthesis did provide evidence that vancomycin-loaded DAC®-coated plates and screws not only are associated with a significant reduction of infection but also protect from the occurrence of septic nonunion, compared to uncoated implants [58]. This study is the very first demonstration that bone healing in a contaminated fracture can be improved by using an osteosynthesis coated with a fast-resorbable, antibiotic-loaded hydrogel.

7. Clinical results and applications

The DAC® hydrogel received the CE mark at the end of year 2013. The available kit (www.coatingdac.com or www.dac-coating.com) is composed of a prefilled syringe, containing 300 mg sterile DAC® powder, that is filled at surgery with a solution of 5 mL sterile water for injection, eventually mixed with the desired antibiotic(s); this allows to obtain, in approximately 3–5 min, the antibiotic-loaded hydrogel, at a DAC® concentration of 6% (w/v) and at an antibiotic concentration usually ranging from 20 mg/mL to 50 mg/mL, depending on the surgeon's choice. The surgeons can choose the antibiotic from among a list of antibacterials previously tested as being compatible with the hydrogel (Novagenit SRL, data on file). A few minutes after reconstitution, the hydrogel can be directly spread onto the implant, which is then inserted into the body in the usual way. If necessary, once reconstituted, the hydrogel may remain at ambient temperature for up to 4 h.

Two large multicenter, randomized, prospective clinical trials were undertaken in Europe, within the 7th European Framework Programme (project # 277988), funded by the European Commission.

In a first trial, a total of 380 patients, scheduled to undergo primary (n = 270) or revision (n = 110) total hip (N = 298) or knee (N = 82) joint replacement with cementless or hybrid (partially cemented) implants were included [59, 60]. The patients were randomly assigned, in six European orthopedic centers, to receive an implant either with the DAC® coating, intraoperatively loaded with antibiotics (treatment group), or without the coating (control group). Pre- and postoperative assessment of clinical scores, wound healing, laboratory tests and X-ray exams were performed at fixed time intervals. Overall, 373 patients were available at a mean follow-up of 14.5 ± 5.5 months (range 6–24). On average, a volume of 8.3 mL hydrogel was used to coat an implant. The most often used antibiotics were vancomycin and gentamicin at a concentration of 5% and 3.2%, respectively. Fifteen patients received an implant with a combined vancomycin and meropenem antibiotic coating; four patients received an implant coated with teicoplanin 5% or ceftazidime 5% or amphotericin B 5%, all in a second-stage procedure for previous infection. Eleven surgical site infections were observed in the control group and only one in the treatment group (6% vs. 0.6%; P = 0.003). No local or systemic side effects related to the DAC® hydrogel coating were reported and no detectable interference with implant osteointegration was noted.

In the other multicenter, prospective study, 256 patients, undergoing osteosynthesis for a closed fracture, were randomly assigned, in 5 European orthopedic centers, to receive the antibiotic-loaded DAC® coating or to a control group, without coating. At a mean follow-up of

Reference	Number of patients	Follow-up (months)	DAC-treated SSI rate (%)	Controls SSI rate (%)	P	Side effects
Romanò et al. [58]	380	14.5 ± 5.5	0.6	6	0.003	None
Malizos et al. [60]	256	18.1 ± 4.5	0	4.6	<0.02	None

SSI, surgical site infection.

Table 1. Summary of the main results of the published clinical multicenter trials on DAC® hydrogel coating in orthopedics and trauma.

18.1 ± 4.5 months (range 12–30), 253 patients were available for evaluation. On average, 5.7 mL (range: 1–10 mL) of DAC® hydrogel was needed to coat the implant. Gentamicin and vancomycin were the most used antibiotics, at concentration of, respectively, 4% or 2%. Six surgical site infections (4.6%) were observed in the control group compared to none in the treated group ($P < 0.02$). No local or systemic side effects related to DAC® hydrogel coating were observed and no detectable interference with bone healing was reported [61, 62] (cf. **Table 1**).

Preliminary results of the possible use of the DAC® hydrogel coating in one-stage exchange of infected prosthesis did also recently show the efficacy and safety of the device in this challenging application [63]. Further studies are currently under way concerning joint replacement in bone tumors, spine surgery, exposed fractures and dentistry.

8. Economic impact

Periprosthetic joint infections (PJI) are associated with increased costs for public health systems, mainly due to additional surgeries, prolonged hospitalization, increased length of rehabilitation and increased use of antibiotics [64]. Moreover, PJIs are associated with an increase in morbidity and mortality [65]. Unless novel, effective measures are taken to reduce the incidence of surgical site infections (SSIs), these complications will become an accruing burden to the health care system in the next two decades [66, 67].

Cost-effectiveness of antibacterial coatings of joint prostheses can be calculated, comparing their direct and indirect hospital costs with those of unprotected implants, taking into consideration the expected surgical site infection rate and using a decision-analytic modeling approach, as previously described by Diaz-Ledezma et al. [68] and Kapadia et al. [69].

Table 2 reports an algorithm used to calculate the overall economic impact of DAC® hydrogel coating. The following variables are included for calculation: average cost and number of primary joint replacements; average cost of the antibacterial coating per patient; incidence of PJI and expected reduction of infection rate with the use of the coating; average cost of PJI treatment and expected number of cases.

Various scenarios can be simulated with the reported algorithm, depending on the relative value given to each variable.

Considering the undiscounted price of DAC® hydrogel at our institution of € 585.00 per package and two packages of DAC® as the standard use per patient, it can be calculated that, if the

	Without ABC		With ABC
Joint replacement, average cost per patient		a	
Number of joint replacements/year		b	
Total cost of joint replacements/year		c = a*b	
ABC, cost per patient	0 (zero)		d
% of expected PJI		e	
% reduction of PJI with ABC		f	
Expected number of infections	g = b*(e/100)		h = b*(e/100)*(1-f/100)
PJI treatment, cost per case		i	
Costs for all septic complication treatments/year	j = g*i		k = h*i
Overall costs for joint replacement incl. septic complications/year	l = c+j		m = c+k
Total costs for ABC			n = b*d
Total costs	o = l		p = m+n
Balance (Medical costs with ABC/without ABC)		q = o-p	
% Balance (Medical costs with ABC/without ABC)		q' = (p/o)*100	

Table 2. Algorithm used to estimate the economical impact of antibacterial coating technologies.

	Without ABC		ABC
Joint replacement, average cost per patient		8,000 €	
Number of joint replacements/year		1,000	
Total cost of joint replacements/year		8,000,000 €	
ABC, cost per patient	0 €		**1,170 €**
% of expected PJI		2.6	
% reduction of PJI with ABC		90	
Expected number of infections	26.0		2.6
PJI treatment, cost per case		50,000 €	
Costs per all septic complication treatment/year	1,300,000 €		130,000 €
Costs for joint replacement incl. septic complications/year	9,300,000 €		8,130,000 €
Total costs for ABC			1,170,000 €
Total costs	**9,300,000 €**		**9,303,000 €**
Balance		**0 €**	
% Balance		**100.0%**	

Table 3. In this simulation, assuming an average cost of primary joint replacement of € 8000 per patient, an average cost of DAC® of € 1170 per case (i.e., two packages per patient), an expected reduction of postsurgical infections by using the coating of 90% and an average cost of PJI treatment of € 50,000 [6, 74], it can be calculated that DAC® is in economical balance if used in a population of patients with an expected periprosthetic infection rate, without the coating, of 2.6%.

coating is able to reduce surgical site infection by 90% [60], DAC® is in economical balance if applied to a population of patients with an expected rate of septic complications (without the coating) of 2.6% (**Table 3**).

According to a similar calculation, if applied on a large scale, to a selected population of patients with at least one risk factor for infection and an expected incidence of infection, without the coating, of 5%, DAC® would provide, in a medium size country, like Italy (approximately 160,000 joint replacements per year), annual direct cost savings of approximately € 43,200,000 (or 1080 € per patient). An expected incidence of postsurgical infection of 5% applies to patients with at least one risk factor for infection, which is at least 25% of all patients undergoing joint replacement [70, 71].

The present analysis is very conservative. One package of DAC® is in fact sufficient in the vast majority of primary implants. Secondly, recent studies point out how the long-term average cost of PJI is much higher than € 50,000, largely exceeding € 100,000 per patient [13]; finally, the algorithm does not include indirect costs, like those deriving from treatment complications, functional inability, work loss and compensation, medicolegal costs, increased mortality rate and quality of life reduction.

9. Conclusions

Biofilm- and implant-related infections represent a dramatic and increasing burden worldwide. Available data show that hyaluronic acid has a proven *in vitro* antiadhesive/antibiofilm effect against some of the most common pathogens, and HA has been used safely, alone or in combination with other polymers, with satisfactory results in different conditions associated with biofilm-related chronic infections. Clinical data in various applications, including dentistry, urology, wound management, dermatology and orthopedics, paved the way to the possible use of HA as a protective coating barrier of implants.

The chemical derivatization of hyaluronic acid with polylactic acid allows the formation of graft copolymers, which, when contacted with an aqueous medium, can be used to produce hydrogels, like the recently CE-marked DAC®, with appropriate characteristics for easy preparation and application at the time of surgery. Resulting medicated hydrogel is transparent, easily spreadable over a surface, like a titanium prosthesis, and has a specifically designed duration; moreover, it has proven, peculiar, antiadhesive and antibiofilm capabilities. If required, it may also be easily loaded, at surgery, with antibacterial agents that will be released over the following hours or few days in effective high local concentrations. In fact, as a passive protective barrier, DAC® hydrogel has some limits. Among others, the antiadhesive/antibiofilm effect is limited and may vary, depending on the type of the microorganism, the bacterial load, the local environment, etc.; moreover, HA protection may be neutralized by the possible ability of some bacteria to produce hyaluronidase, an enzyme that catalyzes the degradation of hyaluronic acid [52], while collagen and hyaluronan may even become possible ligands for microbial attachment in particular situations, or the coating can be covered by other host's proteins to which bacteria may anchor [72, 73]. To overcome some of these limits, possible loading of the hyaluronic-based hydrogel with antibiotics is technically feasible and has been found safe in various preclinical and clinical settings, being a possible option for clinicians.

In fact, both *in vitro* and *in vivo* studies did confirm the safety and efficacy of the hydrogel coating with and without loaded antibacterials.

Clinical results also clearly point out the efficacy of the DAC® coating to significantly reduce early postsurgical infection after joint replacement or internal osteosynthesis, without any detectable local side effect both concerning wound and bone healing. Moreover, no changes in organ-specific serum markers or systemic unwanted effects were noted. The high biocompatibility of its basic constituents and the short time (less than 3 days) needed for a complete hydrogel resorption make the possible occurrence of longer term side effects quite unlikely.

Finally, economical consideration points out the high cost-to-benefit ratio of the large-scale use of DAC® coating, especially in a population with at least one risk factor for infection.

The versatility of the device and its safety profile may open the way to application in other surgical fields that share similar infection risk as orthopedics and trauma.

Author details

Gaetano Giammona[1,2], Giuseppe Pitarresi[1], Fabio Salvatore Palumbo[1], Susanna Maraldi[3], Sara Scarponi[3] and Carlo Luca Romanò[3]*

*Address all correspondence to: carlo.romano@grupposandonato.it

1 Department of Biological, Chemical and Pharmaceutical Science and Technologies, Universityà degli Studi di Palermo, Palermo, Italy

2 Institute of Biophysics at Palermo, Italian National Research Council, Palermo, Italy

3 Department of Reconstructive Surgery of Osteo-articular Infections C.R.I.O. Unit, I.R.C.C.S. Galeazzi Orthopaedic Institute, Milano, Italy

References

[1] Roemling U, Balsalobre C. Biofilm infections, their resilience to therapy and innovative treatment strategies. Journal of Internal Medicine. 2012;**272**:541-561

[2] Romanò CL, Romanò D, Logoluso N, Drago L. Bone and joint infections in adults: A comprehensive classification proposal. European Orthopaedics and Traumatology. 2011; **1**(6):207-217

[3] Cats-Baril W, Gehrke T, Huff K, Kendoff D, Maltenfort M, Parvizi J. International consensus on periprosthetic joint infection: Description of the consensus process. Clinical Orthopaedics and Related Research. 2013;**471**:4065-4075

[4] Drago L, Lidgren L, Bottinelli E, Villafañe JH, Berjano P, Banfi G, Romanò CL, Sculco TP. Mapping of microbiological procedures by the members of the International Society of Orthopaedic Centers (ISOC) for periprosthetic infections diagnosis. Journal of Clinical Microbiology. 2016;**2016**:1402-1403. Pii: JCM.00155-16

[5] Kamath AF, Ong KL, Lau E, Chan V, Vail TP, Rubash HE, et al. Quantifying the burden of revision total joint arthroplasty for periprosthetic infection. The Journal of Arthroplasty. 2015;**30**(9):1492-1497

[6] Romanò CL, Romanò D, Logoluso N, Meani E. Septic versus aseptic hip revision: How different ? Journal of Orthopaedics and Traumatology. 2010;**11**(3):167-174

[7] Parisi TJ, Konopka JF, Bedair HS. What is the long-term economic societal effect of periprosthetic infections after THA? A Markov analysis. Clinical Orthopaedics and Related Research. 2017 Jul;**475**(7):1891-1900

[8] Bonnevialle P, Bonnomet F, Philippe R, Loubignac F, Rubens-Duval B, Talbi A, Le Gall C, Adam P. SOFCOT. Early surgical site infection in adult appendicular skeleton trauma surgery: A multicenter prospective series. Orthopaedics & Traumatology, Surgery & Research. 2012 Oct;**98**(6):684-689

[9] Berbari EF, Osmon DR, Lahr B, Eckel-Passow JE, Tsaras G, Hanssen AD, Mabry T, Steckelberg J, Thompson R. The Mayo prosthetic joint infection risk score: Implication for surgical site infection reporting and risk stratification. Infection Control and Hospital Epidemiology. 2012;**33**:774-781

[10] Heppert V Acute infection after osteosynthesis. European Instructional Lectures. Vol. 12, 2012, 13th EFORT Congress, Berlin, Germany. Springer Science & Business Media Ed. 2012. pp. 25-31. ISBN: 3642272932, 9783642272936

[11] Keene DJ, Mistry D, Nam J, Tutton E, Handley R, Morgan L, Roberts E, Gray B, Briggs A, Lall R, Chesser TJ, Pallister I, Lamb SE, Willett K. The ankle injury management (AIM) trial: A pragmatic, multicentre, equivalence randomised controlled trial and economic evaluation comparing close contact casting with open surgical reduction and internal fixation in the treatment of unstable ankle fractures in patients aged over 60 years. Health Technology Assessment. 2016 Oct;**20**(75):1-158

[12] Oliveira PR, Carvalho VC, da Silva Felix C, de Paula AP, Santos-Silva J, Lima AL. The incidence and microbiological profile of surgical site infections following internal fixation of closed and open fractures. Revista Brasileira de Ortopedia. 2016 Feb 2;**51**(4):396-399

[13] Poultsides LA, Liaropoulos LL, Malizos KN. The socioeconomic impact of musculoskeletal infections. The Journal of Bone and Joint Surgery. American Volume. 2010;**92**:e13

[14] Gristina AG, Naylor P, Myrvik Q. Infections from biomaterials and implants: A race for the surface. Medical Progress through Technology. 1988;**14**(3-4):205-224

[15] Gristina AG, Shibata Y, Giridhar G, Kreger A, Myrvik QN. The glycocalyx, biofilm, microbes, and resistant infection. Seminars in Arthroplasty. 1994;**5**(4):160-170

[16] Dastgheyb S, Parvizi J, Shapiro IM, Hickok NJ, Otto M. Effect of biofilms on recalcitrance of staphylococcal joint infection to antibiotic treatment. Journal of Infectious Diseases. 2015;**211**:641-650

[17] Busscher HJ, van der Mei HC, Subbiahdoss G, Jutte PC, van den Dungen JJ, Zaat SA, et al. Biomaterial-associated infection: Locating the finish line in the race for the surface. Science Translational Medicine. 2012;**4**(153):153rv10

[18] Romanò CL, Scarponi S, Gallazzi E, Romanò D, Drago L. Antibacterial coating of implants in orthopaedics and trauma: A classification proposal in an evolving panorama. Journal of Orthopaedic Surgery and Research. 2015;**10**:157

[19] Moriarty TF, Grainger DW, Richards RG. Challenges in linking preclinical anti-microbial research strategies with clinical outcomes for device-associated infections. European Cells & Materials. 2014;**28**:112-128

[20] Leach JB, Schmidt CE. Hyaluronan. Encyclopedia of Biomaterials and Biomedical Engineering. New York: Marcel Dekker; 2004. pp. 779-789

[21] Liao YH, Jones SA, Forbes B, Martin GP, Brown MB. Hyaluronan: Pharmaceutical characterization and drug delivery. Drug Delivery. 2005;**12**:327-342

[22] Volpi N, Schiller J, Stern R, Solt_es L. Role, metabolism, chemical modifications and applications of hyaluronan. Current Medicinal Chemistry. 2009;**16**:1718-1745

[23] Ardizzoni A, Neglia RG, Baschieri MC, Cermelli C, Caratozzolo M, Righi E, et al. Influence of hyaluronic acid on bacterial and fungal species, including clinically relevant opportunistic pathogens. Journal of Materials Science. Materials in Medicine. 2011; **22**:2329-2338

[24] Pitarresi G, Palumbo FS, Calascibetta F, Fiorica C, Di Stefano M, Giammona G. Medicated hydrogels of hyaluronic acid derivatives for use in orthopedic field. International Journal of Pharmaceutics. 2013 Jun 5;**449**(1-2):84-94. DOI: 10.1016/j.ijpharm.2013.03.059. Epub 2013 Apr 12

[25] Laurencin C, Lane JM. Poly (lactic acid) and poly (glycolic acid): Orthopedic surgery applications. In: Brighton C, Friedlaender G, Lane JM, editors. Bone Formation and Repair, Rosemont, Am. Acad. Orthop. Surg. 1994. pp. 325-339

[26] Pavesio A, Renier D, Cassinelli C, Morra M. Anti-adhesive surfaces through hyaluronan coatings. Medical Device Technology. 1997 Sep;**8**(7):20-21 24-7

[27] Morra M, Cassineli C. Non-fouling properties of polysaccharide-coated surfaces. Journal of Biomaterials Science. Polymer Edition. 1999;**10**(10):1107-1124

[28] Cassinelli C, Morra M, Pavesio A, Renier D. Evaluation of interfacial properties of hyaluronan coated poly(methylmethacrylate) intraocular lenses. Journal of Biomaterials Science. Polymer Edition. 2000;**11**(9):961-977

[29] Kadry AA, Fouda SI, Shibl AM, Abu El-Asrar AA. Impact of slime dispersants and anti-adhesives on in vitro biofilm formation of Staphylococcus Epidermidis on intraocular

lenses and on antibiotic activities. The Journal of Antimicrobial Chemotherapy. 2009 Mar;**63**(3):480-484

[30] Drago L, Cappelletti L, De Vecchi E, Pignataro L, Torretta S, Mattina R. Antiadhesive and antibiofilm activity of hyaluronic acid against bacteria responsible for respiratory tract infections. APMIS. 2014 Oct;**122**(10):1013-1019

[31] Marcuzzo AV, Tofanelli M, Boscolo Nata F, Gatto A, Tirelli G. Hyaluronate effect on bacterial biofilm in ENT district infections: A review. APMIS. 2017 Sep;**125**(9):763-772

[32] Pirnazar P, Wolinsky L, Nachnani S, Haake S, Pilloni A, Bernard GW. Bacteriostatic effects of hyaluronic acid. Journal of Periodontology. 1999;**70**:370-374

[33] Carlson GA, Dragoo JL, Samimi B, Bruckner DA, Bernard GW, Hedrick M, Benhaim P. Bacteriostatic properties of biomatrices against common orthopaedic pathogens. Biochemical and Biophysical Research Communications. 2004 Aug 20;**321**(2):472-478

[34] Radaeva IF, Kostina GA, Il'ina SG, Kostyleva RN. Antimicrobial activity of hyaluronic acid. Zhurnal Mikrobiologii, Epidemiologii, i Immunobiologii. Jan-Feb 2001;(1):74-75

[35] Harris LG, Richards RG. *Staphylococcus aureus* adhesion to different treated titanium surfaces. Journal of Materials Science. Materials in Medicine. 2004 Apr;**15**(4):311-314

[36] Junter GA, Thébault P, Lebrun L. Polysaccharide-based antibiofilm surfaces. Acta Biomaterialia. 2016 Jan;**30**:13-25

[37] Romanò CL, De Vecchi E, Bortolin M, Morelli I, Drago L. Hyaluronic acid and its composites as a local antimicrobial/antiadhesive barrier. Journal of Bone and Joint Infection. 2017;**2**(1):63-72

[38] Torretta S, Marchisio P, Rinaldi V, Gaffuri M, Pascariello C, Drago L, Baggi E, Pignataro L. Topical administration of hyaluronic acid in children with recurrent or chronic middle ear inflammations. International Journal of Immunopathology and Pharmacology. 2016 Sep;**29**(3):438-442

[39] Damiano R, Cicione A. The role of sodium hyaluronate and sodium chondroitin sulphate in the management of bladder disease. Therapeutic Advances in Urology. 2011 Oct;**3**(5):223-232

[40] Constantinides C, Manousakas T, Nikolopoulos P, Stanitsas A, Haritopoulos K, Giannopoulos A. Prevention of recurrent bacterial cystitis by intravesical administration of hyaluronic acid: A pilot study. BJU International. 2004;**93**(9):1262-1266

[41] Shao Y, Shen ZJ, Rui WB, Zhou WL. Intravesical instillation of hyaluronic acid prolonged the effect of bladder hydrodistention in patients with severe interstitial cystitis. Urology. 2010;**75**(3):547-550

[42] Lipovac M, Kurz C, Reithmayr F, Verhoeven HC, Huber JC, Imhof M. Prevention of recurrent bacterial urinary tract infections by intravesical instillation of hyaluronic acid. International Journal of Gynaecology and Obstetrics. 2007;**96**:192-195

[43] Damiano R, Quarto G, Bava I, Ucciero G, Palumbo MI, Autorino R, et al. Prevention of recurrent urinary tract infections by intravesical administration of hyaluronic acid and chondroitin sulphate: A placebo-controlled randomised trial. European Urology. 2011;**59**:645-651

[44] Ciani O, Arendsen E, Romancik M, Lunik R, Costantini E, Di Biase M, Morgia G, Fragalà E, Roman T, Bernat M, Guazzoni G, Tarricone R, Lazzeri M. Intravesical administration of combined hyaluronic acid (HA) and chondroitin sulfate (CS) for the treatment of female recurrent urinary tract infections: A European multicentre nested case-control study. BMJ Open. 2016 Mar 31;**6**(3):e009669

[45] Johannsen A, Tellefsen M, Wikesjö U, Johannsen G. Local delivery of hyaluronan as an adjunct to scaling and root planing in the treatment of chronic periodontitis. Journal of Periodontology. 2009 Sep;**80**(9):1493-1497

[46] Sapna N, Vandana KL. Evaluation of hyaluronan gel (Gengigel(®)) as a topical applicant in the treatment of gingivitis. Journal of Investigative and Clinical Dentistry. 2011 Aug;**2**(3):162-170

[47] Eick S, Renatus A, Heinicke M, Pfister W, Stratul SI, Jentsch H. Hyaluronic acid as an adjunct after scaling and root planing: A prospective randomized clinical trial. Journal of Periodontology. 2013 Jul;**84**(7):941-949

[48] Drago L, Boot W, Dimas K, Malizos K, Hänsch GM, Stuyck J, Gawlitta D, Romanò CL. Does implant coating with antibacterial-loaded hydrogel reduce bacterial colonization and biofilm formation in vitro ? Clinical Orthopaedics and Related Research. 2014 Nov;**472**(11):3311-3323

[49] Giammona G, Pitarresi G, Palumbo FS, Romanò CL, Meani E, Cremascoli E. Antibacterial hydrogel and use thereof in orthopedics. 2010. WO 2010/086421 A1.

[50] Pitarresi G, Palumbo FS, Albanese A, Fiorica C, Picone P, Giammona G. Self assembled amphiphilic hyaluronic acid graft copolymers for targeted release of antitumoral drug. Journal of Drug Targeting. 2010;**18**:264-276

[51] Pitarresi G, Palumbo FS, Fiorica C, Calascibetta F, Di Stefano M, Giammona G. Injectable in situ forming microgels of hyaluronic acid-g-polylactic acid for methylprednisolone release. European Polymer Journal. 2013;**49**:718-725

[52] Palumbo FS, Pitarresi G, Mandracchia D, Tripodo G, Giammona G. New graft copolymers of hyaluronic acid and polylactic acid: Synthesis and characterization. Carbohydrate Polymers. 2006;**66**:379-385

[53] Cloutier M, Mantovani D, Rosei F. Antibacterial coatings: Challenges, perspectives, and opportunities. Trends in Biotechnology. 2015 Nov;**33**(11):637-652

[54] Hynes WL, Walton SL. Hyaluronidases of gram-positive bacteria. FEMS Microbiology Letters. 2000;**183**:201-207

[55] Giavaresi G, Meani E, Sartori M, Ferrari A, Bellini D, Sacchetta AC, Meraner J, Sambri A, Vocale C, Sambri V, Fini M, Romanò CL. Efficacy of antibacterial-loaded coating in an

in vivo model of acutely highly contaminated implant. International Orthopaedics. 2014 Jul;**38**(7):1505-1512

[56] Boot W, Vogely HC, Nikkels PGJ, Pouran B, van Rijen M, Dhert WJA, Gawlitta D. Local prophylaxis of implant-related infections using a hydrogel as carrier. European Cells and Materials. 2015;**30**(2):19

[57] Boot W, Gawlitta D, Nikkels PGJ, Pouran B, van Rijen MHP, Dhert WJA, Vogely HC. Hyaluronic acid-based hydrogel coating does not affect bone apposition at the implant surface in a rabbit model. Clinical Orthopaedics and Related Research. 2017 Jul;**475**(7):1911-1919

[58] Lovati AB, Drago L, Bottagisio M, Bongio M, Ferrario M, Perego S, Sansoni V, De Vecchi E, Romanò CL. Systemic and local administration of antimicrobial and cell therapies to prevent methicillin-resistant *Staphylococcus epidermidis*-induced femoral nonunions in a rat model. Mediators of Inflammation. 2016;**2016**:9595706. DOI: 10.1155/2016/9595706

[59] Malizos K, Scarponi S, Simon K, Blauth M, Romanò C. Clinical results of an anti-bacterial hydrogel coating of implants: A multi-centre, prospective, comparative study. Bone & Joint Journal. 2015, 2015;**97-B**(16):138

[60] Romanò CL, Malizos K, Capuano N, Mezzoprete R, D'Arienzo M, Van Der Straeten C, Scarponi S, Drago L. Does an antibiotic-loaded hydrogel coating reduce early post-surgical infection after joint arthroplasty? Journal of Bone and Joint Infection. 2016;**1**:34-41

[61] Logoluso N, Malizos K, Blauth M, Danita A, Simon K, Romanò C. Anti-bacterial hydrogel coating of osteosynthesis implants: Early clinical results from a multi-center prospective trial. European Cells and Materials. 2015;**30**(2):35

[62] Malizos K, Blauth M, Danita A, Capuano N, Mezzoprete R, Logoluso N, Drago L, Romanò CL. Fast-resorbable antibiotic-loaded hydrogel coating to reduce post-surgical infection after internal osteosynthesis: A multicenter randomized controlled trial. Journal of Orthopaedics and Traumatology. 2017 Jun;**18**(2):159-169. DOI: 10.1007/s10195-017-0442-2

[63] Gallazzi E, Capuano N, Scarponi S, Morelli I, Romanò CL. Does one-stage exchange with antibacterial coating of implants provide similar results to a two-stage procedure for the treatment of peri-prosthetic joint infection? Bone & Joint Journal. 2017;**99**(Suppl. 1): 19-19

[64] Parvizi J, Pawasarat IM, Azzam KA, Joshi A, Hansen EN, Bozic KJ. Periprosthetic joint infection: The economic impact of methicillin-resistant infections. The Journal of Arthroplasty. 2010;**25**(6 Suppl):103-107

[65] Berend KR, Lombardi AV Jr, Morris MJ, Bergeson AG, Adams JB, Sneller MA. Two-stage treatment of hip periprosthetic joint infection is associated with a high rate of infection control but high mortality. Clinical Orthopaedics and Related Research. 2013; **471**(2):510-518

[66] Kurtz S, Ong K, Lau E, et al. Projections of primary and revision hip and knee arthroplasty in the United States from 2005 to 2030. The Journal of Bone and Joint Surgery. American Volume. 2007;**89**(4):780

[67] Kurtz SM, Ong KL, Schmier J, et al. Future clinical and economic impact of revision total hip and knee arthroplasty. The Journal of Bone and Joint Surgery. American Volume. 2007;**89**(Suppl 3):144

[68] Diaz-Ledezma C, Lichstein PM, Dolan JG, Parvizi J. Diagnosis of periprosthetic joint infection in medicare patients: Multicriteria decision analysis. Clinical Orthopaedics and Related Research. 2014;**472**(11):3275-3284

[69] Kapadia BH, Johnson AJ, Issa K, Mont MA. Economic evaluation of chlorhexidine cloths on healthcare costs due to surgical site infections following total knee arthroplasty. The Journal of Arthroplasty. 2013 Aug;**28**(7):1061-1065

[70] Bozic KJ, Ong K, Lau E, et al. Estimating risk in Medicare patients with THA: An electronic risk calculator for periprosthetic joint infection and mortality. Clinical Orthopaedics and Related Research. 2013;**471**(2):574-583

[71] Eka A, Chen AF. Patient-related medical risk factors for periprosthetic joint infection of the hip and knee. Annals of Translational Medicine. 2015;**3**(16):233

[72] Barton AJ, Sagers RD, Pitt WG. Bacterial adhesion to orthopedic implant polymers. Journal of Biomedical Materials Research. 1996 Mar;**30**(3):403-410

[73] Birkenhauer E, Neethirajan S, Weese JS. Collagen and hyaluronan at wound sites influence early polymicrobial biofilm adhesive events. BMC Microbiology. 2014 Jul 16;**14**:191

[74] Garrido-Gómez J, Arrabal-Polo MA, Girón-Prieto MS, Cabello-Salas J, Torres-Barroso J, Parra-Ruiz J. Descriptive analysis of the economic costs of periprosthetic joint infection of the knee for the public health system of Andalusia. Journal of Arthroplasty. 2013;**28**(7):1057-1060